T0332943

B

Suresh H. Moolgavkar
Editor

Scientific Issues in Quantitative Cancer Risk Assessment

With 48 Illustrations

Birkhäuser
Boston · Basel · Berlin

Suresh H. Moolgavkar
Fred Hutchinson Cancer Research Center
Division of Public Health Sciences
Seattle, WA 98104
USA

Library of Congress Cataloging-in-Publication Data
Scientific issues in quantitative cancer risk assessment/Suresh
    Moolgavkar, editor.
        p.   cm.
    Based on the 11th conference held in 1989 at Snowbird, Utah,
sponsored by SIMS.
    Includes bibliographical references.
    ISBN 0-8176-3501-7
    1. Carcinogenesis—Mathematical models—Congresses. 2. Health
risk assessment—Statistical methods—Congresses. I. Moolgavkar,
Suresh H. II. SIAM Institute for Mathematics and Society.
    [DNLM: 1. Carcinogens—congresses. 2. Cell Transformation.
Neoplastic—etiology—congresses. 3. Risk—congresses. QZ 202
S416 1989]
    RC268.5.S365   1990
    616.99'4071—dc20
    DNLM/DLC   90-55

Printed on acid-free paper.

ISBN 0-8176-3501-7
ISBN 3-7643-3501-7

Camera-ready copy supplied by the editor.
Printed and bound by Edwards Brothers, Inc., Ann Arbor, Michigan.
Printed in the U.S.A.

9 8 7 6 5 4 3 2 1

Foreword

In 1974, the Societal Institute of the Mathematical Sciences (SIMS) initiated a series of five-day Research Application Conferences (RAC's) at Alta, Utah, for the purpose of probing in depth societal fields in light of their receptivity to mathematical and statistical analysis. The first 10 conferences addressed ecosystems, epidemiology, energy, environmental health, time series and ecological processes, energy and health, energy conversion and fluid mechanics, environmental epidemiology: risk assessment, atomic bomb survivor data: utilization and analysis, and modern statistical methods in chronic disease epidemiology.

These *Proceedings* are a result of the eleventh conference of Scientific Issues in Quantitative Cancer Risk Assessment, which was held in 1989 at Snowbird, Utah. For five days, 45 speakers and observers contributed their expertise in the relevant biology and statistics. The presentations were timely and the discussion was both enlightening and at times spirited. This volume hopefully presents a record that will be useful both now and in the future.

Suresh H. Moolgavkar of the Fred Hutchinson Cancer Research Center (Seattle, WA) and the University of Washington (Seattle) chaired the Conference.

The Conference was supported by the following organizations:

The Dow Chemical Company
Haskell Laboratory, E.I. duPont
Health and Welfare Canada
International Life Sciences Institute
Monsanto Company
National Cancer Institute
Procter & Gamble
Shell Oil Company
U.S. Department of Energy
U.S. Environmental Protection Agency
U.S. Food and Drug Administration

SIMS is grateful to the above organizations for their support of this Conference.

D.L. Thomsen, Jr.
President, SIMS

Preface

That society must regulate exposures to potential human carcinogens is axiomatic. That such regulation is sometimes based on guesswork rather than realistic scientific assessment of risk is an inevitable consequence of our ignorance of the carcinogenic process. However, things are changing. The last decade has seen great strides in our understanding of possible molecular mechanisms underlying carcinogenesis, and "oncogene" and "tumor suppressor gene" have become, if not household words, basic concepts in the biology of cancer. The summer of 1989 seemed like a good time to bring together scientists working in disciplines relevant to cancer risk assessment to exchange ideas on how science might be applied to the problem of quantifying risk in populations exposed to putative carcinogens. The conference was not about risk assessment per se but about the scientific questions that must be addressed before rational risk assessment is possible. The chapters in this book represent the proceedings of the conference.

In the current paradigm, somatic mutations, broadly defined as heritable changes in the genome of a cell, play a central role in carcinogenesis. The chapters by Clifton, Knudson, Mendelsohn, and Wiseman discuss different aspects of the role of mutations in carcinogenesis. Clifton suggests that cancers are derived from small subpopulations of clonogenic cells, and that initiation is too common an event to be a locus-specific mutation in the traditional sense. Mendelsohn briefly discusses the role of somatic mutations in carcinogenesis and reviews methods for measuring locus-specific mutations in human populations. Knudson discusses the roles of oncogenes and tumor suppressor genes within the framework of a two-event model for carcinogenesis. Wiseman reviews the experimental literature on oncogene activation in spontaneous and chemically induced liver and lung tumors in rodents.

It is now widely appreciated that cancer risks depend to a large extent on the dynamics of tissue growth and differentiation. Somatic mutation rates are influenced by the rates of cell division and the size of susceptible subpopulations of cells depends on the relative rates of division and differentiation. A large number of papers presented at the conference deal with the relative roles of mutations and cell proliferation in determining cancer risk. The chapters by Pitot et al. and Schwarz et al. assess the role of cell proliferation in rodent hepatocarcinogenesis experiments. Cohen and Ellwein, Moolgavkar et al., and Chen and Moini discuss the relative roles of mutations and cell proliferation within the framework of a two-event model for carcinogenesis. Grosser and Whittemore present a statistical analysis of lung adenomas in mice exposed to urethane with particular emphasis on the role of urethane in adenoma growth. Krewski and Murdoch discuss cancer modeling with intermittent exposures and introduces the concept of equivalent constant dose for the multistage and the two-event models. The chapter by Yamasaki and Fitzgerald discusses the role of altered gap junctional intercellular communication in carcinogenesis.

Individuals show a great deal of variation in their response to carcinogenic agents. The biological basis of such interindividual variation is discussed by Harris. Incorporation of interindividual variation into biologically based mathematical models for carcinogenesis has substantial impact on the conclusions. This is discussed in the chapter by Portier.

Finally, whenever possible, assessment of risk to humans should be based on human epidemiologic data. Becher and Wahrendorf discuss a simple measure of risk in human populations, the unit risk, and illustrate its usefulness by computing unit risks associated with exposure to arsenic and benzene. Kaldor discusses the carcinogenic potential of drugs used for cancer chemotherapy and points out that studies of second cancers following chemotherapy provide unique opportunities for understanding quantitative aspects of human carcinogenesis.

The last morning of the conference was devoted entirely to a free-ranging discussion of issues raised during previous sessions. Wilson's summary captures some of the spirit of that discussion.

The excellent talks, the lively discussion, and the idyllic setting all contributed to the success of the conference.

Seattle, Washington   Suresh H. Moolgavkar

Contents

Contributors

Heiko Becher
Department of Epidemiology, German Cancer Research Center, D-6900 Heidelberg 1, Federal Republic of Germany

Albrecht Buchmann
German Cancer Research Center, Institute of Biochemistry, D-6900 Heidelberg 1, Federal Republic of Germany

Harold Campbell
McArdle Laboratory, University of Wisconsin, Madison, WI 53706, USA

Chao W. Chen
U.S. Environmental Protection Agency (RD-689), Washington, D.C. 20460, USA

Kelly H. Clifton
K4/330 University of Wisconsin, Clinical Cancer Center, Madison, WI 53792, USA

Samuel M. Cohen
Department of Pathology & Microbiology, University of Nebraska Medical Center, Omaha, NE 68105, USA

Mathisca de Gunst
Department of Mathematics, Free University, 1081 HV Amsterdam, The Netherlands

Leon B. Ellwein
University of Nebraska Medical Center, Omaha, NE 68105, USA

D. James Fitzgerald
International Agency for Research on Cancer, 69371 Lyon Cedex 08, France

Stella Grosser
University of California at Los Angeles, School of Public Health/Biostatistics, Center for the Health Sciences, Los Angeles, CA 90024, USA

Curtis C. Harris
National Cancer Institute, Building 37, Room 2C05, Bethesda, MD 20892, USA

James R. Hully
McArdle Laboratory, University of Wisconsin, Madison, WI 53706, USA

J. Kaldor
International Agency for for Research on Cancer, Unit of Biostatistics Research and Informatics, 69372 Lyon Cedex 08, France

Alfred G. Knudson, Jr.
Fox Chase Cancer Center, Institute for Cancer Research, Philadelphia, PA 19111, USA

Daniel Krewski
Biostatistics & Computer Applications Division, Health and Welfare Canada, Environmental Health Centre, Ottawa, Ontario K1A OL2, Canada

Werner Kunz
German Cancer Research Center, Institute of Biochemistry, D-6900 Heidelberg 1, Federal Republic of Germany

Georg Luebeck
Fred Hutchinson Cancer Research Center, Division of Public Health Sciences, Seattle, WA 98104, USA

Mortimer L. Mendelsohn
Director, Biomedical and Environmental Research Program, Lawrence Livermore National Laboratory, L-452, Livermore, CA 94550, USA

Assad Moini
U.S. Environmental Protection Agency (RD-689), Washington, D.C. 20460, USA

Suresh H. Moolgavkar
Fred Hutchinson Cancer Research Center, Division of Public Health Sciences, Seattle, WA 98104, USA

Duncan J. Murdoch
Department of Statistics and Actuarial Science, University of Waterloo,
Waterloo, Ontario N2L 3G1, Canada

Mark J. Neveu
McArdle Laboratory, University of Wisconsin, Madison, WI 53706, USA

Henry C. Pitot
McArdle Laboratory, University of Wisconsin, Madison, WI 53706, USA

Christopher J. Portier
National Institute of Environmental Health Sciences, Biometry and Risk
Assessment Program, Research Triangle Park, NC 27709, USA

Tahir A. Rizvi
McArdle Laboratory, University of Wisconsin, Madison, WI 53706, USA

Larry W. Robertson
University of Kentucky, Graduate Center for Toxicology, Lexington, KY
40506, USA

Michael Schwarz
German Cancer Research Center, Institute of Biochemistry, D-6900
Heidelberg 1, Federal Republic of Germany

Jürgen Wahrendorf
Department of Epidemiology, German Cancer Research Center, D-6900
Heidelberg 1, Federal Republic of Germany

Alice S. Whittemore
Stanford University School of Medicine, Department of Health Research and
Policy, Palo Alto, CA 94305-5092, USA

James D. Wilson
American Industrial Health Council, Washington, D.C. 20037, USA

Roger W. Wiseman
National Institute of Environmental Health Sciences, Research Triangle Park,
NC 27709, USA

Hiroshi Yamasaki
Programme of Multistage Carcinogenesis, Unit of Mechanisms of
Carcinogenesis, International Agency for Research on Cancer, 69371 Lyon
Cedex 08, France

The Clonogenic Cells of the Rat Mammary and Thyroid Glands: Their Biology, Frequency of Initiation, and Promotion/Progression to Cancer

Kelly H. Clifton

ABSTRACT: Quantitative transplantation and hormonal manipulation methods have been developed for studies of radiation carcinogenesis in the rat mammary and thyroid glands. Experiments with these systems have led to the conclusions a) that cancers are derived from small subpopulations of clonogenic cells in both organs, and b) that initiation, the first step in carcinogenesis, is a common event at the cellular level. It is postulated that any intracellular event which increases the probability of mutation in future rounds of cell division may serve as initiation. Several radiation-induced non-mutational intracellular processes are now recognized. In addition to mutation in growth-regulating genes, we suggest that the induction of one or more non-mutational processes may serve as initiation.

1. INTRODUCTION

The results of experiments based on quantitative transplantation of rat thyroid and mammary cells have led to two conclusions of significance to those interested in the cell population kinetics of these two important organs and to those concerned with the sequence of events in the carcinogenic process. Firstly, the epithelia of both glands contain subpopulations of cells which are clonogenic and are the apparent progenitor cells of carcinomas. Secondly, initiation, the first step in the carcinogenic process, is a common event in these clonogenic cells. It is the purpose of this discussion to review the evidence for these two principal conclusions, and to comment on some of their corollaries. In the following, the term *clonogen* is used operationally to denote a cell which when transplanted to syngeneic host and exposed to appropriate physiological stimulation will proliferate and differentiate to form a multicellular clonal glandular structure. The term *initiation* is used operationally to denote the first change in a cell, carcinogen-induced or of unknown cause, which leads after promotion/progression to an overt cancer. It is of importance to note that both operational definitions are based on observations of the capacity of a cell to give rise to a clonal multicellular structure, normal or malignant. It follows that a graft inoculum contained at least one clonogen if a clonal glandular structure developed in the graft site in response to hormonal stimulation. Likewise, it follows that some of the cells which had been exposed to an agent were initiated if cancer later occurred in higher freqency and/or with shorter latency in grafts which contained exposed cells than in those which contained unexposed cells. Neither normal nor initiated clonogens can yet be defined by morphological or biochemical criteria.

2. THE BIOLOGY OF CLONOGENS

Research in my laboratory on the biology of clonogenic cells was stimulated by the thought that it is likely that not all epithelial cells in a tissue are susceptible to the induction of cancer. Rather it seemed to us that those cells necessary for repair and repopulation after tissue injury or extreme physiological stress are the most likely cancer precursor cells. If this is true, a corollary is that biochemical or other analyses of whole tissue cell populations might well not reflect the situation in that critical cell subpopulation from which cancers can arise.

We chose to look for evidence of such a cell subpopulation in rat mammary tissue. Rat mammary glands, like those of humans, are susceptible to radiation carcinogenesis [17,48], the glands and tumors derived from them are subcutaneous and readily palpable, gland pieces are readily transplantable in syngeneic animals [48], and much is known of the hormonal regulation of normal and neoplastic mammary growth [12,3]. Mammary glands were removed from normal donor rats and minced, and their cells were enzymatically monodispersed. Known cell numbers were then inoculated into the gland-free interscapular white fat pads (i.s. pads) of syngeneic recipients. The recipient rats were co-grafted with an *MtT* (mammotropic pituitary tumor) to serve as an indwelling source of high levels of *Prl* (prolactin), the prime hypophyseal stimulant of mammary growth and differentiation [8]. The aim was to design the experimental system to maximize the likelihood that any cell in the graft inoculum which was capable of division and differentiation would do so.

Three weeks after mammary cell transplantation, many of the graft sites contained free-standing spherical mammary alveolar structures which were filled with milk (Fig. 1) [14]. This observation, and the results of experiments which followed, encouraged investigation of the thyroid gland which, like the mammary gland, is susceptible both to hormonal control of function and proliferation, and to radiogenic cancer in both rats and humans [19]. When dispersed thyroid cells were grafted into the i.s. pads of thyroidectomized recipients, some of the grafted cells responded to the elevated *TSH* (thyroid stimulating hormone) levels by giving rise to multicellular follicles which formed colloid and showed evidence of thyroid hormone secretion (Fig. 2) [6,18].

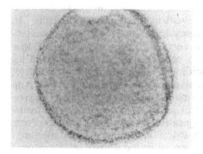 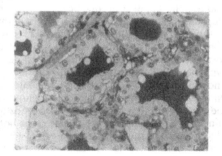

FIGURE 1. Whole mount preparation of a simple alveolar unit (AU) in an interscapular fat pad that had been injected with mammary cells 3.5 weeks earlier. The recipient rat was grafted with a Prl-secreting pituitary tumor, X785 [26].

FIGURE 2. Section through follicles which developed in an interscapular fat pad which had been inoculated 4 weeks previously with thyroid cells. The recipient rat was thyroidectomized, X260 [42].

Terminal dilution transplantation bioassays were developed using the mammary alveolar structures or *AU* (alveolar units) and thyroid follicular structures or *FU* (follicular units) as the endpoints [27,18]. In both systems, serial dilutions of enzymatically monodispersed cells are inoculated in up to five graft sites in the subcutaneous fat pads per recipient rat. Some of the grafted mammary cells are stimulated to form AU by high Prl levels secreted by co-grafts of MtT, and some of the grafted thyroid cells form FU in response to the elevated TSH levels induced by thyroidectomy of the recipients. Three or four weeks later, the grafted fat pads are removed and examined for the presence or absence of AU or FU in the graft sites. The data are analyzed as described below.

The results of studies with these two systems lead to the conclusion that both glands contain subpopulations of clonogenic epithelial cells which give rise to AU or FU and to cancer. Evidence for this conclusion is derived from morphological studies during AU and FU formation, the relationship between the inoculated cell number and the fraction of graft sites with AU or FU, the response to radiation exposure, and cell population kinetics in glands *in situ*. These data are discussed below.

Morphological evidence: If cell suspensions are prepared directly by enzymatic digestion of whole young adult female mammary glands and transplanted to MtT-bearing recipients, the *AD50* (alveolar dose-50%) is $\sim 2 \times 10^3$ cells, i.e. on average, $\sim 2 \times 10^3$ cells are required per graft site to give rise to at least one AU in 50% of the sites [27]. If comparable suspensions are prepared from young male thyroid glands and grafted in thyroidectomized recipients, the *FD50* (follicular dose-50%) is $\sim 10^2$ cells per graft site [18,43].

Morphological studies at intervals after mammary [26] or thyroid [42] cell transplantation revealed no evidence of cell reaggregation. The sequence of events was similar after grafting of either mammary or thyroid cells. Extensive inflammation was present one day after grafting. The inflammation was subsiding during days two to four, and scattered single epithelial-appearing cells were observed. By days four to five, small spheres comprised of two to four cells each were first seen along with scattered single cells (Fig. 3). By days six to eight, mammary graft sites contained hollow multicellular spheres; thyroid graft sites contained small follicles with evidence of colloid secretion. Three to four weeks after grafting, mammary graft sites contained milk-filled alveolar spheres lined with normal-appearing secretory epithelium accompanied by myoepithelial cells; thyroid graft sites contained thyroid follicles lined with stimulated normal-appearing thyroid epithelial cells and filled with colloid (Fig. 4). These AU and FU were 0.3-0.7 mm in diameter, and readily detectable with a dissecting microscope.

Although both AU and FU appear from histological studies to arise from single hormonally-stimulated cells, in terms of the total numbers of inoculated cells, it is conceivable that AU and/or FU could arise by cell reaggregation after inoculation of the suspensions. Indeed, such reaggregation of thyroid cells has been observed in culture under the influence of TSH [36,40]. However, studies of the loss of radioactivity from sites of inoculation of prelabelled cells show that the actual numbers of presumably proliferation-competent cells which remain in the graft sites are considerably less than the numbers of cells inoculated. Mammary and thyroid cells were labelled *in situ* in donor rats during hormone-induced glandular hyperplasia by several injections of the DNA precursor tritiated thymidine. Mammary cells from the labelled glands were then grafted in MtT-bearing recipient rats, and cells from labelled thyroid glands were grafted in thyroidectomized rats. Four days after grafting, only ~ 1-2% of the radioactivity in the inoculated mammary cells remained in the graft sites, and five days after grafting, only ~ 7-11% of the radioactivity in the inoculated thyroid cells remained

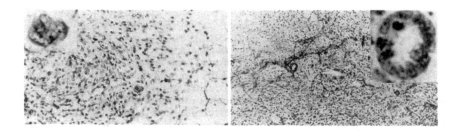

FIGURE 3. Sections through graft sites in MtT-bearing rats which had been inoculated with mammary cells. Left: Five days after inoculation, X1630. Insert of developing structure, X4225. Right: Six days after inoculation, X845. Insert of developing alveolus, X5850 [26].

[26,45]. Thus, the numbers of proliferation-competent cells available for reaggregation are a small fraction of those inoculated.

Two additional observations are more consistent with a clonal origin than a reaggregation origin of AU and FU. Firstly, in the absence of elevated tropic hormone levels in the graft recipients, few or no AU or FU are formed from grafted cells. Four weeks after grafting of thyroid cells into euthyroid recipients, however, elevation of TSH by thyroidectomy of the recipients led to FU formation with the same frequency per grafted cell as when the TSH levels were elevated from the time of grafting (Fig. 5) [45], indicating that the FU-forming cells persisted at the graft sites. It seems very unlikely that migration and reaggregation of scattered single epithelial cells could easily occur in the by then well healed and uninflamed graft sites four weeks after transplantation.

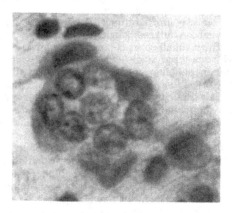

FIGURE 4. Solid organized epithelial structure common on days 4 and 5 after grafting of thyroid cells in thyroidectomized recipients, X1100 [42].

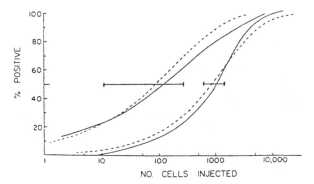

FIGURE 5. Effect of thyroidectomy of the recipient rats on FD50 assay curves. "% positive" indicates percentage of graft sites with FU. Dashed curve to left, assay curve in recipients thyroidectomized one day before grafting, and to right in intact recipients. Both groups were killed for observation 28 days after grafting. Solid curve to left, assay curve for recipients that were thyroidectomized 28 days after grafting, and to right in intact recipients. Both groups were killed for observation 56 days after grafting. Horizontal bars are 95% confidence limits of the FD50 values in the latter groups [45].

Secondly, when sufficient numbers of cells are inoculated to produce a few AU or FU per site, these tend to be separate, of approximately the same size as in sites with smaller cell numbers, and of similar morphology. Often, they will be found immediately adjacent to one another with no evidence of coalescence. In sites grafted with cell numbers in great excess of the AD50 or FD50, i.e. at cell numbers in which groups of clonogens would be expected to lie close together in the small graft volume, AU and FU occur in clusters (Figs. 6 and 7). If AU and FU arose by reaggregation of cells, one would expect that larger units would be formed at the outset from grafts of large cell numbers.

Quantitative transplantation evidence: The numbers and concentrations of clonogens are calculated from the transplantation bioassay data according to the model of Porter et al. [47] using the following relationships:

$$P = 1 - \exp (-M)$$

and:

$$\log M = \log K + S \log Z$$

where P is the probability that an AU or FU will develop in a given graft site, M is the mean number of clonogens per site in that concentration of graft inoculum, K is the clonogenic fraction, Z is the mean total number of cells per site in that concentration of graft inoculum, and S is the slope of the relationship [9]. S and K are estimated with the aid of a maximum likelihood iterative procedure.

According to this model, if the structures scored as the end points are monoclonal in origin and there are no cell interactions during their formation, S will equal or approximate 1.0. If two cells are required to form a structure, S will approximate 0.5, and if more cells are necessary, S is progressively smaller [27,30]. In several tens of AD50 assays performed in our laboratories during more than a decade by several different individuals, S has routinely approximated 1.0, usually falling within the range of 0.9 to 1.1 and including 1.0 within its 95% confidence limits [27]. In FD50 assays, S tends to vary in the range of 0.75-1.05, suggesting interactions, probably hormonal in nature, of the grafted cells with the host (*vide infra*) [50]. When performed with careful technique, the assays are remarkably reproducible. For example, the clonogen concentrations in the mammary glands of young adult females of the W/Fu and F344 rat strains were assayed using grafts

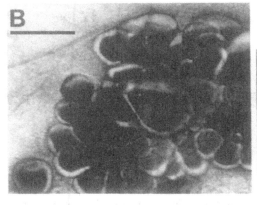
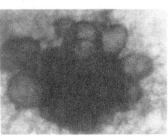

FIGURE 6. Cluster of AU in a whole mount of a site grafted with a large mammary clonogen number. Bar is 1.0 μm [32].

FIGURE 7. Cluster of FU in site grafted with a large number of thyroid clonogens.

of two different strain-specific MtT lines as the Prl sources in the recipients. The resultant AD50 values were 704 and 786 cells per graft site, respectively, and plots of the positive graft sites against the cell dose per site in the two strains were virtually superimposable (Kamiya, Gould and Clifton, unpublished). The reproducibility of repeated assays of normal thyroid clonogen concentrations is illustrated in Table 1. If AU and FU arose by reaggregation, one would expect that the slope S would be <0.5, i.e. would be a complex inverse function of the number of cells required to reaggregate to form a recognizeable structure, and that assay results would be less predictable.

These data strongly support the conclusion that the individual glandular structures which develop in grafts of monodispersed mammary and thyroid epithelial cells in response to appropriate hormonal stimulation are derived by proliferation and differentiation of individual clonogenic cells.

Evidence from radiation studies: Clonogen concentrations have been determined in both glands by assays performed immediately following exposure to graded doses of x-rays or gamma rays *in vivo* or *in vitro*, and the clonogen survival has been calculated by the ratio of the AD50 or FD50 values of the unirradiated control cells to those of the irradiated cells (Fig. 8). The resultant radiation dose-clonogen survival curves, like those of cells irradiated and assayed by clonal colony formation in monolayer culture systems, are characterized by an initial shoulder region with a low cell death rate at low radiation doses followed by

Table 1.   Reproducibility of transplantation assay of normal thyroid clonogens [50].

Investigator	Number of Assays	Slope, S	Clonogenic Fraction, K
KMG	5	0.802	0.0120
FED	3	0.794	0.0133
	geometric means:	0.799	0.0125
	95% conf. lim	0.730, 0.868	0.0061, 0.0255

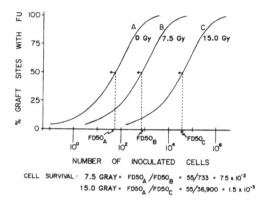

FIGURE 8. Sample calculation of thyroid clonogen survival from FD50 assay results.

exponentially decreasing survival with increasing dose. These survival curves based on FD50 and AD50 assays have been analyzed according to the multitarget-single hit model [20]: $\qquad S = 1 - \{1 - \exp(-D/D_o)\}^n$
where S is the fractional cell survival, D is the radiation dose, D_o is the inverse of the terminal slope of the curve, and n is the extrapolation number, i.e. the value at the intercept of the terminal slope extrapolated to D = 0 Gy. In theory, D_o reflects intracellular target sensitivity, and n reflects target number [20]. Following exposure to sparsely ionizing photon radiations, the D_o and n values so derived (Table 2) are within the range seen in clonal colony forming assays of mammalian cell survival [28]. Furthermore, as in culture systems, the dose-survival curves from similar AD50 assays after exposure of rats to graded doses of neutrons have little or no initial shoulder (Table 2); the value of n approximates 1.0 and the curve is exponential from a dose of 0 Gy [39].

Alternatively, the cell survival data may be analyzed according to the linear-quadratic model, i.e. $S = \exp -(\alpha D + \beta D^2)$ in which αD represents the intracellular damage resulting from single ionizing events and βD^2 reflects the damage requiring two ionizing events [28]. The data from the current experiments are somewhat better fit by the multitarget-single hit model which has therefore been used as an emperical data summary aid throughout these studies.

Table 2. Radiation dose-clonogen survival parameters with and without in situ repair (ISR) [9].

Tissue	Radiation	ISR	D_o, Gy	n
mammary	x-rays	no	1.3	5
	x-rays	yes	1.3	17
	neutrons	--*	0.97	1
thyroid	x-rays	no	2.0	3
	x-rays	yes	2.0	10

*D_o and n insignificantly different whether assay is performed immediately following exposure or 24 hours later.

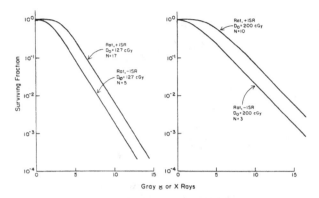

FIGURE 9. Radiation dose-cell survival curves for mammary clonogens (left) and thyroid clonogens (right) illustrating the effect of in situ repair (ISR(. "-ISR" are data from assays performed immediately after exposure; "+ISR" are data from assays performed on cells left in the glands in situ for 24 hours after exposure before removal for assay.

When either gland is irradiated with sparsely ionizing photon radiation and left *in situ* for several hours before removal and dispersion for assay, the initial shoulder of the survival curve is enlarged and the n value is increased three- to five-fold, indicating post-irradiation intracellular repair (Fig. 9) [9]. This *ISR (in situ* repair) is completed within four to six hours after exposure of mammary glands (Fig. 10) [38], and has been found to occur as well in rat hepatocytes as measured in a similar clonogen transplantation assay [9]. Terzaghi-Howe (personal communication) has recently observed that when tracheal epithelial cells are exposed to x-rays *in situ* and removed for culture 2-3 hours later, survival is fourfold that following exposure of the cells in dispersed suspensions. ISR has recently been demonstrated in cultures of mammary ductal fragments (Kamiya, Gould and Clifton, unpublished) and may thus depend on close intercellular contact and/or the intercellular transfer of chemicals. ISR is not seen after exposure of mammary cells to densely ionizing neutrons (Table 2) [38]. The kinetics of ISR are consistent with the kinetics of post-irradiation intracellular repair of potentially lethal damage in cultured cells [28].

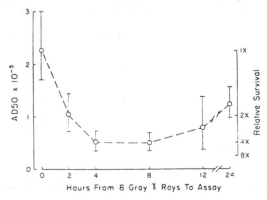

FIGURE 10. Kinetics of in situ repair (ISR) [38].

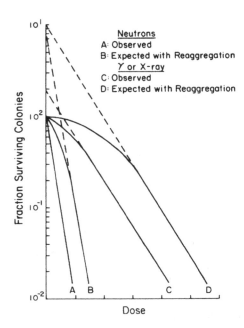

FIGURE 11. Samples of "observed" survival curves after neutron (curve A) or low LET irradiation (curve C) calculated from clonogen assay data. Curves B and D indicate what would be expected if the AU or FU assay endpoints were the result of cell reaggregation.

Were AU and FU derived by reaggregation of several to many cells following transplantation, the initial survival curve shoulder would be expected to be very large, i.e. the n value would be expected to exceed the number of cells required to aggregate to form a scoreable structure (Fig. 11). Furthermore, the n value would be expected to be >1.0 following exposure to densely ionizing neutrons as well as after exposure to sparsely ionizing photon radiation. In analyses according to the alternative linear-quadratic model, the linear term would be very small, and in some cases would even be negative -- on biological grounds, an unlikely finding. All of these data and considerations are thus most consistent with the monoclonal origin of AU and FU.

Evidence from studies of cell population kinetics and differentiation control: Recent studies of the physiological control of clonogen numbers in both glands *in situ* show a separation of the control of the major cell populations and the small subpopulations of assayable clonogens. For example, when young adult female rats were exposed for 48 days either to elevated Prl levels alone *(Prl+)* or to Prl+ coupled with glucocorticoid deficiency *(Prl+/Glc-)*, the total numbers of recoverable dispersed mammary cells increased about sevenfold in both experimental groups; recoverable cell numbers did not change significantly in untreated age control rats (Table 3) [33]. AD50 assays revealed that the subpopulations of clonogens in the mammary glands of Prl+ rats did not increase during the experimental period despite the marked increase in total cells (Table 3). In contrast, in Prl+/Glc- rats the clonogenic subpopulations increased more than five fold.

The findings above may be attributed to the fact that Prl+ in the presence of Glc leads to secretory, probably terminal, differentiation of daughter cells of

Table 3.   Effect of 48 days of exposure to Prl+/Glc- or Prl+ on total DNA, tritiated thymidine incorporation, recoverable cell numbers and total clonogens in the rat mammary gland [33].

Mammary Donors		DNA		Cells	Clonogens
Age, days	Treatment	Total μg	DPM/mg[a]	$x10^{-6b}$	$x10^{-3b}$
50-55	untreated	82 ± 12	17 ± 3	5.39	1.75
98-103	untreated	87 ± 12	20 ± 4	6.02	1.75
98-103	Prl+/Glc-	626 ± 388	63 ± 24	42.0	8.92
98-103	Prl+	ND	ND	39.1	1.36

[a]Radioactive disintegrations per minute from tritiated thymidine incorporated into DNA.
[b]Total recoverable cells and assayable clonogens

clonogens. In the Prl+/Glc-rats, this differentiative pathway is blocked [13,5]. In recent experiments, we have demonstrated that Prl+ and/or elevated estrogen levels lead to the apparently clonal development of small non-secretory branched ductal units (DU) in grafts of mammary cells in adrenogonadectomized recipients (Fig. 12) [32]. The addition of Glc results instead in AU formation. The addition of progesterone promotes DU formation and antagonizes the action of Glc. The evidence suggests, but does not prove, that the formation of AU and DU are alternative differentiative pathways of the same mammary cells [32].

The thyroid glands of rats treated with the thyroid peroxidase inhibitor aminotriazole (ATA) enter a rapid proliferative phase which, by six weeks of treatment results in goiters containing 35-40 fold more DNA than is in the normal gland. This goitrogenesis is in response to the elevation in TSH levels which occurs because of the decrease in thyroid hormone titers which result in turn from the inhibition of the enzyme which is essential to thyroxine synthesis. Further goitrous growth slows markedly thereafter despite continued high TSH titers. During the first six week period of rapid growth, the total subpopulation of clonogenic cells per gland remains unchanged; total clonogens per gland increase rapidly by fivefold during the second six week period (Fig. 13). If the goitrogenic ATA is withdrawn after 12 weeks, the goiters involute with a loss in total DNA; the total population of clonogens remains elevated during involution, however (Groch and Clifton, unpublished). These data further confirm that the thyroid epithelial cell population is comprised of two subpopulations, i.e. a large subpopulation of cells with a limited proliferative potential and a small subpopulation of clonogenic cells with a considerably greater proliferative potential.

If rats are given an iodine deficient diet and are treated with $KClO_4$, which blocks the uptake of iodine by the thyroid follicular cells and hence blocks thyroxine synthesis, goitrous hyperplasia ensues as with ATA treatment. However, goitrous growth is not followed by an expansion of the subpopulation of clonogens during the second six week period of such treatment as it is during ATA treatment (Table 4) (Groch and Clifton, unpublished). Thus the subpopulation of clonogenic cells behaves differently from the remainder of the epithelial cell population in response to the two goitrogenic regimens; furthermore, the two treatments induce different responses among the clonogens following the period of rapid goitrous growth.

The difference in the response of the clonogens to the two treatments cannot be attributed to differences in the TSH levels. Radioimmunoassays at

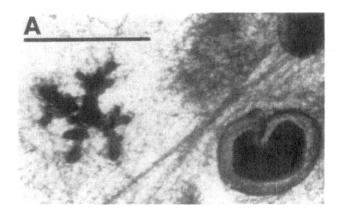

FIGURE 12. Comparison of a ductal unit (DU to left) with an AU (to right) in a whole mount of a mammary cell graft site. Bar is 1.0 μm [32].

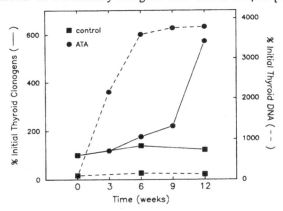

FIGURE 13. Effect of ATA treatment (circles) on total thyroid DNA (dashed line) and total clonogen numbers per thyroid gland (solid line). Squares indicate untreated control values (Groch and Clifton, unpublished).

Table 4.   Effect of two goitrogenic regimens on serum TSH titers, total thyroid DNA, and total assayable clonogens per thyroid gland (K.M. Groch and K.H. Clifton, unpublished).

Treatment Regimen	Weeks	TSH ng/ml[c]	μg DNA per Gland[c]	Clonogens per Gland
untreated	---	2.0±0.2	34±4	52,100
ATA[a]	12	22.2±5.5	968±41	241,900
$KClO_4$ + LID[b]	12	26.7±0.5	449±37	21,500

[a]Aminotriazole, 0.05% in drinking water.
[b]0.5% $KClO_4$ drinking water plus low iodine diet
[c]mean ± standard error

intervals during treatment with each goitrogenic regime show comparable and greatly elevated serum TSH titers (Table 4). Furthermore, clonogens prepared from glands of rats treated with ATA or $KClO_4$ and low iodine diet respond similarly on transplantation to clonogens from normal thyroid glands, i.e. there is no apparent change in their sensitivity to the elevated TSH levels and/or other conditions in thyroidectomized graft recipients (Groch and Clifton, unpublished).

In summary, both the quantitative and qualitative data indicate the existence of clonogenic epithelial cell subpopulations in both the mammary and thyroid glands which are quantifiable by transplantation assay. The data are generally inconsistent with the alternative suggestion that the multicellular glandular structures which arise in the transplantation sites are the result of reaggregation of several to many cells.

3. NEOPLASTIC INITIATION AND HORMONAL PROMOTION

We undertook these studies to investigate the process of radiation carcinogenesis at the cellular level *in vivo* using overt neoplasia as the end point. The first experiments were aimed at measuring the frequency of neoplastic initiation, the first step in carcinogenesis, per target cell. In the ideal experiment, each initiated cell would give rise to an overt carcinoma. To approximate this ideal, our experiments were designed to hormonally maximize promotion/progression after irradiation as described below.

Experimental designs: In practice, mammary or thyroid cells were irradiated *in vivo* before enzymatic dispersion or *in vitro* after dispersion. Inocula containing known numbers of surviving clonogenic cells calculated from the radiation dose-cell survival studies were then grafted in the subcutaneous fat pads of syngeneic recipient rats. To achieve chronic Prl+ to stimulate mammary clonogen proliferation, a piutitary gland and a capsule containing estrone were implanted in the spleen of each mammary cell recipient [16]. Each recipient was also adrenalectomized and maintained on minimal mineralocorticoid replacement therapy to produce chronic Glc- and block secretory differentiation of the grafted mammary cells [7]. This combination of Prl+/Glc- is the most effective endocrine promoting/progressing regimen for mammary cells of which we are aware; indeed, given time, it alone can lead to mammary carcinoma development [16].

To achieve chronically elevated TSH levels (TSH+) to stimulate the proliferation of grafted thyroid cells, the recipient rats were thyroidectomized and maintained on an iodine deficient diet [6]. As noted below, recent data show that the TSH+ initially induced by this regimen decreases with time in some recipients during long term experiments.

The recipient rats were inspected frequently for the development of tumors. Gross and microscopic examination revealed that the tumors found at the graft sites had developed within the glandular structures derived from the transplanted clonogens. Only those tumors diagnosed as carcinomas on histopathological examination are included in the data analyses.

The frequency of neoplastic initiation: Carcinomas developed in 50% of the sites grafted with an average of ~ 120 surviving mammary clonogens that had been exposed to 7 Gy (700 rad) gamma radiation *in situ* in the donor rats one day before transplantation (Fig.14) [15]. When corrected for the small carcinoma incidence in the sites grafted with unirradiated cells, this corresponds to one carcinoma per ~ 280 irradiated clonogens or a radiogenic initiation probability of ~ 3.6×10^{-3} per clonogen (Table 5) [10]. The time of maximum cancer hazard was ~ 600 days after grafting.

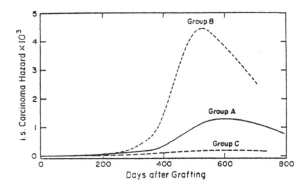

FIGURE 14. Mammary cancer hazard in grafts of known numbers of surviving clonogens plotted against time. Group A: graft sites inoculated with ~120 surviving 7 Gy gamma-irradiated mammary clonogens. Groups B and C: each site inoculated with ~1800 unirradiated mammary clonogens. Transplant sites of Group B were irradiated with 7 Gy x-rays 35 days after grafting [15].

In another experiment, aliquots of monodispersed thyroid cells were mock irradiated or exposed to 5 Gy 250 kVp x-rays *in vitro*, and serial dilutions containing ~26 to ~411 surviving thyroid clonogens were transplanted per graft site. TSH+ was induced in the recipients as described above. The highest incidence of cancer per clonogen occured in those sites which had received the lowest irradiated cell dose per graft [44]. The net radiation-related incidence in the sites grafted with ~26 clonogens was one cancer per ~440 clonogens, or a radiogenic initiation probability of ~2.3 x 10^{-3} per clonogen. Interestingly, over the sixteenfold difference in irradiated clonogen numbers per graft site, there was but a fourfold difference in total carcinoma frequency, i.e. the efficiency of promotion/progression of radiation-initiated clonogens to cancer was inversely related to the number of clonogens grafted per site. In the sites which received unirradiated cells, there was no clear relationship between the number of cells grafted per site and the cancer incidence, i.e. similar numbers of cancers occured in each of the experimental groups [44].

In a more recent experiment, serially diluted suspensions containing ~11 to ~720 surviving 5 Gy-irradiated thyroid clonogens were grafted per site [50]. As in the first experiment, the carcinoma incidence per clonogen was inversely related to the number of irradiated clonogens grafted per site without added unirradiated cells (Fig. 15). The incidence in the group which had received ~720 clonogens per site was but 2.3 times greater than in those sites which received ~11 clonogens. In the latter group, the net incidence was one cancer per ~75 irradiated clonogens or an initiation risk of ~1.3 x 10^{-2} per clonogen (Table 5) [10].

Cell interactions during promotion/progression: In the thyroid carcinogenesis experiment described above, there were additional groups of recipients that received the same numbers of irradiated clonogens, but these were mixed with 1300 to 2500 unirradiated clonogens per graft site [50]. The addition of the unirradiated thyroid clonogens to the irradiated clonogen inocula resulted in suppression of carcinoma incidence (Fig. 15). The decrease in cancer incidence ranged from a fraction of -0.16 to -0.35 as compared to the incidence in grafts of irradiated cells alone, and was proportional to the ratio of unirradiated clonogens to irradiated clonogens in the inocula [50].

14

Table 5.   Frequencies of neoplastic initiation per grafted clonogen [15,50].

Clonogen type:	mammary	thyroid
Radiation dose:	7 Gy gamma	5 Gy x-rays
Number of graft sites:	131	82
Surviving clonogens		
per graft site:	120	11
group total:	1.6×10^4	9.0×10^2
Number of carcinomas:	65	28
Carcinoma risk per		
clonogen:	4.1×10^{-3}	3.1×10^{-2}
Ratio, cancer:clonogens		
uncorrected:	1:240	1:32
corrected*:	1:280	1:75

*for background incidence

The cancer-suppressing effect of increasing cell numbers per graft site appears to result at least in part from reestablishment of the thyroid-pituitary feedback system mediated by thyroid hormone production by the FU which develop in the grafts. Radioimmunoassays were performed 44 weeks after transplantation on sera from rats which had been grafted with 180 clonogens in each of five sites (900 clonogens per rat), and in those which had been similarly grafted with a total of 25 clonogens per rat. Serum TSH titers were ~20 times the normal values in rats grafted with a total of ~25 clonogens, regardless of whether they had been maintained on a normal or an iodine deficient diet (Table 6) [50]. In contrast, serum TSH was within the normal range in rats which had been grafted with ~900 clonogens and were subsequently maintained on a normal diet. TSH levels in comparable rats maintained on an iodine deficient diet were significantly higher,

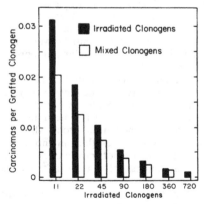

FIGURE 15. Effect of thyroid clonogen number per graft site on the efficiency per clonogen of promotion/progression to cancer. Solid bars: cancer incidence per clonogen in grafts of the indicated numbers of surviving 5 Gy-irradiated clonogens per site. Open bars: cancer incidence per clonogen in grafts which had received the indicated number of irradiated clonogens mixed with 1260-2520 unirradiated clonogens [50].

Table 6. Effect of total thyroid clonogen numbers grafted per animal on serum TSH 44 weeks after transplantation to thyroidectomized rats fed normal (ND) or low iodine diet (LID) [50].

Number of Rats	Clonogens Grafted per Rat	Diet	Serum TSH ng/ml + s.d.
17	25	LID	43+32
10	25	ND	50+21
16	900	LID	5.3+2.6
6	900	ND	2.0+0.8

about twice that of rats fed a normal diet, but were much less than in either group of rats which had received the low cell dose. Hence, the cells in those sites in thyroidectomized rats which were grafted with small cell numbers in the carcinogenesis experiments were subjected to more continuous and intensive promotion by TSH+ than were cells in grafts of large cell numbers. Experiments are in progress to determine whether locally acting paracrine or autocrine factors are also involved in the suppression of the promotion/progression of initiated thyroid cells.

Carcinogenesis in situ and clonogen number and proliferation: In the experiment on radiogenic carcinoma incidence per grafted mammary clonogen described above, the cancer incidence was higher in grafts of mammary cells irradiated with 7 Gy x-rays to the graft sites 30 days after transplantation and 48 days after the beginning of Prl+/Glc- than in those exposed before grafting (Fig. 14). This result could be related to amplification of the clonogen population and/or the elevation of the proliferative rate in the gland caused by exposure to Prl+/Glc- as noted above. Consideration of these and other data led to the hypothesis that, all other things being equal, radiogenic cancer incidence may be a direct function, and cancer latency an indirect function, of the number of clonogenic cells present at the time of irradiation [4].

To test this hypothesis, groups of rats were adrenalectomized and implanted intrasplenically with a pituitary gland and estrone capsule on experimental day 0. Groups designated 4X and 4N had been exposed to 0.4 Gy x-rays or 0.1 Gy fission neutrons respectively on day -1 before surgery, and Group C served as unirradiated controls (Table 7) [33]. Groups 2X and 2N were exposed to 0.4 Gy x-rays and 0.1 Gy neutrons respectively on day +48 after amplification of the mammary clonogen subpopulation and during a time when studies of tritiated thymidine incorporation showed that mammary cell proliferation was elevated about fourfold above normal. All animals were followed for mammary tumor development *in situ.*

Mammary carcinogenesis was significantly potentiated in those groups irradiated 48 days after the beginning of hormonal stimulation of the clonogens, i.e., as compared to the data from the control group C, the first mammary carcinoma latencies were reduced, and total carcinomas per rat-day were increased, in groups 2X and 2N (Table 7). In accord with the hypothesis, carcinoma latency was significantly greater and total cancer frequency per rat-day was significantly less in group 4X which was x-irradiated on day -1 before the establishment of Prl+/Glc- than in group 2X; both parameters in group 4X were insignificantly different from those of group C (Table 7). In contrast, carcinoma latencies and frequencies in neutron-irradiated group 4N were insignificantly different from those in group 2N (Table 7). Thus the carcinogenic response to a relatively low dose

Table 7.   Effect of radiation type and time of exposure on mammary carcinogenesis in Prl+/Glc- rats* [33].

Group	Irradiation Day	Dose**	Median Days to First Carcinoma	Total Carcinomas per 10^3 Rat Days
C	-	--	467	3.3
2X	+48	40 cGy x	378	5.4
4X	-1	40 cGy x	474	2.8
2N	+48	10 cGy n	367	6.3
4N	-1	10 cGy n	384	4.5

*All rats received an intrasplenic implant of a pituitary gland and estrone capsule on day 0 of the experiment.
**x indicates x-rays; n indicates neutrons.

of x-rays was markedly influenced by the number and/or proliferation rate of the clonogenic target cells at the time of exposure, while the response to neutrons was not. Indeed, the proliferative state of the target cells at the time of exposure to sparsely ionizing gamma or x-rays may be as important as the total target cell number [33].

Terzaghi-Howe (personal communication) has found marked differences in the pre-neoplastic response to neutrons as compared to x-rays. The expression of pre-neoplastic alterations in tracheal epithelial cells induced by x-rays is less in cells exposed and retained in vivo than in cells exposed in suspension and maintained in culture. The pre-neoplastic response of such cells to neutrons is the same under both conditions. The expression of x-ray-induced damage is thus reduced by processes which occur in tissue, while that of neutrons is not.

4. CONCLUSIONS AND IMPLICATIONS

These studies, we believe, well support the hypothesis that the target cells from which mammary and thyroid cancers arise are members of small subpopulations of the epithelial cells of these glands. Members of these subpopulations are capable of forming multicellular clonal glandular structures in response to appropriate hormonal stimulation, i.e. are clonogenic. These clonogens are most likely the cells responsible for repair and repopulation following tissue injury. Furthermore, neoplastic initiation of these clonogens by radiation is a frequent event at the cellular level.

We have discussed the postulate that there is a small but finite probability of a change toward neoplasia with each round of cell division due to intracellular synthetic error or to low levels of carcinogens of external or internal origin [4]. When one clonogenic cell has accumulated changes sufficient in nature and number, overt neoplasia may ensue. This is not a new concept, having been stated in one form or another by Furth [25], Noble [46] and others [2,11].

During the lifespan of the normal rat, the probability of incremental neoplastic change per cell division is sufficiently low, the populations of clonogenic target cells sufficiently small, and the time between mitoses sufficiently long that the achievement of mammary or thyroid cancer is a rare event. According to the reasoning above, if the numbers of clonogens are increased or the times between mitoses are reduced by hormonal conditions that block differentiation and/or stimulate mitoses, the probability of incremental neoplastic change and overt cancer is increased. Any agent which either directly induces incremental neoplastic change, or increases the probability in future cell generations of one or more such incremental neoplastic changes, would thus be expected both to increase the incidence of cancer and to shorten cancer latency.

The high frequency of neoplastic initiation in the clonogens following irradiation described herein is not unique to our studies. Mondal and Heidelberger [41] treated isolated single immortalized mouse fibroblasts with methylcholanthrene and then cultured the cells in the absence of carcinogen. One hundred percent of the clones derived from single treated cells gave rise to morphologically transformed cellular foci, and the latter formed tumors when transplanted to immune-suppressed mice. Thus the methylcholanthrene had brought about preneoplastic change in every exposed cell. Fernandez et al. [24] later described the probabilities of three necessary stages in the neoplastic process from immortalized cell to malignant cell in this system.

In vitro transformation systems have been widely investigated as models of neoplastic initiation, and new combined *in vivo-in vitro* systems have yielded important results. Kennedy [35] has reviewed her studies on *in vitro* transformation as well as the work of others, and has concluded that initiation is a frequent event at the cellular level. Furthermore, Kennedy et al. [34] demonstrated reversal of *in vitro* transformation by post-irradiation treatment with protease inhibitors. Terzhagi-Howe (personal communication) recently found that two to three percent of rat tracheal epithelial cells expressed pre-neoplastic changes after exposure to 0.2 Gy neutrons or 4 Gy x-rays. She and her associates had previously described the disappearance of carcinogen treated pre-neoplastic tracheal epithelial cells when they were placed in close contact with normal tracheal epithelium [49]. Studies of aberrant colony formation in grafts and cultures of mouse mammary cells following irradiation [1,21] have also indicated a high neoplastic initiation frequency.

There are important consequences to the observation that neoplastic initiation as defined here is a common cellular event. Either such initiation is not attributable primarily to mutation in a single specific growth-regulating gene, or the specific locus involved is orders of magnitude more mutable than direct studies of mutation frequencies in mammalian cells have yet revealed. Alternatively, it is possible that a mutational event in any one of a large number of growth-regulating genes or gene copies may constitute initiation. If this be true, the data on the reversibility of initiation must be brought into accord with the concept of the irreversibility of mutation.

We speculate, rather, that initiation is a generic term which includes any events, genetic of epigenetic, which increase the probability of progression of a cell or its progeny to cancer. This includes a) any event which by increasing genomic "noise" increases the probability of the chance appearance of a mutation in one or more of the critical growth-regulating genes during future rounds of cell division, as well as b) immediately-induced mutational events in such loci. According to this view, initiation may include induction of ongoing processes which are either genetic or are epigenetic in the sense that they do not immediately involve irreversible changes in gene sequence.

Kennedy [35] has cited examples of radiation-induced ongoing processes in a variety of experimental systems. Among these is genetic recombination which is induced by radiation in yeast and which does not occur until the exposed cells are mated with unirradiated cells; furthermore, the recombination so induced is between unirradiated chromosomes [22]. On-going cellular responses similar to the "SOS" DNA repair system which is induced by radiation in *E. coli* [37] have recently been reported to occur in irradiated mammalian cells [51]. More recently, aberrant DNA methylation has been shown to be induced by radiation in a high proportion of exposed mammalian cells [31]. Feinberg et al. [23] have reported hypomethylation of DNA cytosine in human colon carcinomas and adenomas as compared to adjacent normal mucosa, and have suggested that such changes may be an early step in carcinogenesis. Hardwick et al. [29] have found increased complexity of methylation of the c-*mos* gene in a variety of "normal" tissues from humans who were exposed to radiation from internally deposited radium. They postulate that the changes observed are epigenetic in origin and may involve methylation of adenosine. Altered DNA methylation, through affects on gene expression, could increase the probability of further changes including some which are pre-neoplastic. These phenomena are cited to illustrate that there are processes other than direct mutational events which may begin as a result of carcinogen exposure in a high proportion of exposed cells. One or more such induced common cellular processes may alter the probability of neoplastic genomic change in subsequent cell generations.

In summary, the commonality of initiation as defined in these studies suggests that it includes any process, general or specific, which increases the probability of rare and irreversible preneoplastic change or changes in future cell generations. The probability of the latter is further increased by post-initiation promotion. It follows that the rare events in the multi-stage process of carcinogenesis in most cases occur during promotion; promotion thus is a critical period for action in the cancer prevention.

It is of importance to note that although at the tissue level, the sequence of events in carcinogenesis may appear orderly--e.g., first hyperplasia, then papilloma, and finally carcinoma--at the cellular and intracellular levels, the processes may very well be largely random, occuring in different sequences and perhaps in different combinations. There indeed is likely to be more than one way to skin a cat.

ACKNOWLEDGEMENTS

The author is indebted to his associates, including particularly Dr. Kenji Kamiya, Kevin Groch and Frederick Domann, for the use of previously unpublished data and valued discussions, and to Mrs. Jane Barnes and Mrs. Joan Mitchen for outstanding technical assistance during the course of these studies. This work was supported by National Cancer Institute grant R37-CA13881 and Department of Energy grant DE FG02 87ER60507. The author receives partial salary support from National Cancer Institute grant P30-CA14520 to the University of Wisconsin Clinical Cancer Center, 600 Highland Avenue, Madison, WI 53792.

REFERENCES

1. Adams, L.M., Ethier, S.P., and Ullrich, R.L. Enhanced in vitro proliferation and in vivo tumorigenic potential of mammary epithelium from BALB/c mice exposed in vivo to gamma-radiation and/or 7,12-dimethylbenz(a)anthracene. Cancer Res., 47:4425-4431, 1987.

2. Albanes, D. and Winick, M. Are cell number and cell proliferation risk factors for cancer? JNCI, 80:772-774, 1988.

3. Clifton, K.H. The physiology of endocrine therapy. In: F.F. Becker (ed.), Cancer, A Comprehensive Treatise, Vol. 5, pp. 573-597. New York, NY:Plenum Press, 1977.

4. Clifton, K.H. Cancer risk per clonogenic cell in vivo: Speculation on the

relationship of both cancer incidence and latency to target cell number. In: K. Lapis and S. Eckhardt (eds.), Lectures and Symposia of the 14th International Cancer Congress, Carcinogenesis and Tumour Progression, Vol. 4, pp. 89-99. Basel:Karger, 1987.

5. Clifton, K.H. and Crowley, J. Effects of radiation type and role of glucocorticoids, gonadectomy and thyroidectomy in mammary tumor induction in MtT-grafted rats. Cancer Res., 38:1507-1513, 1978.

6. Clifton, K.H., DeMott, R.K., Mulcahy, R.T., and Gould, M.N. Thyroid gland formation from inocula of monodispersed cells: Early results on quantitation, function, neoplasia and radiation effects. Int. J. Radiat. Oncol., 4:987-990, 1978.

7. Clifton, K.H. and Furth, J. Ductoalveolar growth in mammary glands of adrenogonadectomized male rats bearing mammotropic pituitary tumors. Endocrinology, 66:893-897, 1960.

8. Clifton, K.H. and Furth, J. Mammotropin effects in tumor induction and growth. In: C.H. Li (ed.), Hormonal Proteins and Peptides, Vol. 8, pp. 75-103. New York, NY:Academic Press, 1979.

9. Clifton, K.H. and Gould, M.N. Clonogen transplantation assay of mammary and thyroid epithelial cells. In: C.S. Potten and J.H. Hendry (eds.), Cell Clones: Manual of Mammalian Cell Techniques, pp. 128-138. Edinburgh:Churchill Livingstone, 1985.

10. Clifton, K.H., Kamiya, K., Groch, K.M., and Domann, F.E. Quantitative studies of rat mammary and thyroid clonogens, the presumptive cancer progenitor cells. In: K.H. Chadwick (ed.), Proceedings, Workshop on Cell Transformation Systems Relevant to Radiation-induced Cancer in Man, in press. Bristol, U.K.:IOP Publishing Ptd., 1989.

11. Clifton, K.H. and Meyer, R.K. Mechanism of anterior pituitary tumor induction by estrogen. Anat. Rec., 125:65-81, 1956.

12. Clifton, K.H. and Sridharan, B.N. Endocrine factors and tumor growth. In: F.F. Becker (ed.), Cancer, A Comprehensive Treatise, Vol. 3, pp. 249-285. New York, NY:Plenum Press, 1975.

13. Clifton, K.H., Sridharan, B.N., and Douple, E.B. Mammary carcinogenesis-enhancing effect of adrenalectomy in irradiated rats with pituitary tumor MtT-F4. J. Natl. Cancer Inst., 55:485-487, 1975.

14. Clifton, K.H., Sridharan, B.N., and Gould, M.N. Risk of mammary oncogenesis from exposure to neutrons or gamma rays: Experimental methodology and early findings. In: Biological and Environmental Effects of Low Level Radiation (IAEA-SM-202/211), pp. 205-211. Vienna:International Atomic Energy Agency.

15. Clifton, K.H., Tanner, M.A., and Gould, M.N. Assessment of radiogenic cancer initiation frequency per clonogenic rat mammary cell in vivo. Cancer Res., 46:2390-2395, 1986.

16. Clifton, K.H., Yasukawa-Barnes, J., Tanner, M.A., and Haning, Jr., R.V. Irradiation and prolactin effects on rat mammary carcinogenesis: Intrasplenic pituitary and estrone capsule implants. J. Natl. Cancer Inst. 75:167-175, 1985.

17. Cronkite, E.P., Shellabarger, C.J., Bond, V.P., and Lippincott, S.W. Studies on radiation-induced mammary gland neoplasia in the rat. I. The role of the ovary in the neoplastic response of breast tissue to total- or partial-body x-irradiation. Radiat. Res., 12:81-93, 1960.

18. DeMott, R.K., Mulcahy, R.T., and Clifton, K.H. The survival of thyroid cells following irradiation: A directly generated single-dose-survival

curve. Radiat. Res., 77:395-403, 1979.

19. Dumont, J.E., Malone, J.F., and Van Herle, A.J. Irradiation and Thyroid Disease: Dosimetric, Clinical and Carcinogenic Aspects, Brussels:Commission of the European Communities, 1980.

20. Elkind, M.M. and Sutton, H. Radiation response of mammalian cells grown in culture. I. Repair of x-ray damage in surviving Chinese hamster cells. Radiat. Res., 13:556-593, 1960.

21. Ethier, S.P. and Ullrich, R.L. Detection of ductal dysplasia in mammary outgrowths derived from carcinogen-treated virgin female BALB/c mice. Cancer Res., 42:1753-1760, 1982.

22. Fabre, F. and Roman, H. Genetic evidence for inducibility of recombination competence in yeast. Proc. Natl. Acad. Sci. USA, 74:1667-1671, 1977.

23. Feinberg, A.P., Gehrke, C.W., Kuo, K.C., and Ehrlich, M. Reduced genomic 5-methylcytosine content in human colonic neoplasia. Cancer Res., 48:1159-1161, 1988.

24. Fernandez, A., Mondal, S., and Heidelberger, C. Probabilistic view of the transformation of cultured C3H/10T1/2 mouse embryo fibroblasts by 3-methylcholanthrene. Proc. Natl. Acad. Sci. USA, 77:7272-7276, 1980.

25. Furth, J. Conditioned and autonomous neoplasms: A review. Cancer Res., 13:477-492, 1953.

26. Gould, M.N., Biel, W.F., and Clifton, K.H. Morphological and quantitative studies of gland formation from inocula of monodispersed rat mammary cells. Exper. Cell Res., 107:405-416, 1977.

27. Gould, M.N. and Clifton, K.H. The survival of mammary cells following irradiation in vivo: A directly generated single-dose-survival curve. Radiat. Res., 72:343-352, 1977.

28. Hall, E.J. Radiobiology for the Radiologist, 3rd edn., Philadelphia:J.B. Lippincott, 1988.

29. Hardwick, J.P., Schlenker, R.A., and Huberman, E. Alteration of the c-mos locus in "normal" tissues from humans exposed to radium. Cancer Res., 49:2668-2673, 1989.

30. Hendry, J.H. Mathematical aspects of colony growth, transplantation kinetics and cell survival. In: Cell Clones: Manual of Mammalian Cell Techniques, pp. 1-12. London:Churchill Livingstone, 1985.

31. Kalinich, J.F., Catravas, G.N., and Snyder, S.L. The effect of gamma radiation on DNA methylation. Radiat. Res., 117:185-197, 1989.

32. Kamiya, K., Gould, M.N., and Clifton, K.H. Ductal versus alveolar differentiation of rat mammary clonogens. (submitted), 1989.

33. Kamiya, K., Higgins, P.D., Tanner, M.A., Yokoro, K. and Clifton, K.H. Clonogenic cells and rat mammary cancer: Effects of hormones, x-rays and fission neutrons. Radiat. Res., in press, November 1989.

34. Kennedy, A.R. In: T.J. Slaga (ed.), Mechanisms of Tumor Promotion, Vol. III, pp. 13-55. Boca Raton, FL:CRC Press, 1984.

35. Kennedy, A.R. Evidence that the first step leading to carcinogen-induced malignant transformation is a high frequency, common event. In: J.C. Barrett and R.W. Tennant (eds.), Carcinogenesis, Vol. 9, pp. 355-364. New York:Raven Press, 1985.

36. Lissitzky, S., Fayet, G., Giraud, A., Verrier, B., and Torresani, J. Thyrotrophin-induced aggregation and reorganization into follicles of isolated porcine-thyroid cells. Eur. J. Biochem., 24:88-99, 1971.

37. Little, J.W., Edmiston, S.H., Pacelli, L.Z., and Mount, D.W. Cleavage of

the Escherichia coli lexA protein by the recA protease. Proc. Natl. Acad. Sci. USA, 77:3225-3229, 1980.

38. Mahler, P.A., Gould, M.N., and Clifton, K.H. The kinetics of in situ repair in rat mammary cells. Int. J. Radiat. Biol., 44:443-446, 1983.

39. Mahler, P.A., Gould, M.N., DeLuca, P.M., Pearson, D.W., and Clifton, K.H. Rat mammary cell survival following irradiation with 14.3 MeV neutrons. Radiat. Res., 91:235-242, 1982.

40. Mak, W.W., Errick, J.E., Chan, R.C., Eggo, M.C., and Burrow, G.N. Thyrotropin-induced formation of functional follicles in primary cultures of ovine thyroid cells. Exper. Cell Res., 164:311-322, 1986.

41. Mondal, S. and Heidelberger, C. In vitro malignant transformation by methylcholanthrene of the progeny of single cells derived from C3H mouse prostate. Proc. Natl. Acad. Sci., 65:219-225, 1970.

42. Mulcahy, R.T., DeMott, R.K., and Clifton, K.H. Transplantation of monodispersed rat thyroid cells: Hormonal effects on follicular unit development and morphology. Proc. Soc. Exp. Biol. Med., 163:100-110, 1980.

43. Mulcahy, R.T., Gould, M.N., and Clifton, K.H. The survival of thyroid cells: In vivo irradiation and in situ repair. Radiat. Res., 84:523-528, 1980.

44. Mulcahy, R.T., Gould, M.N., and Clifton, K.H. Radiation initiation of thyroid cancer: A common cellular event. Int. J. Radiat. Biol., 45:419-426, 1984.

45. Mulcahy, R.T., Rose, D.P., Mitchen, J.M., and Clifton, K.H. Hormonal effects on the quantitative transplantation of monodispersed rat thyroid cells. Endocrinology 106:1769-1775, 1980.

46. Noble, R.L. Hormonal control of growth and progression in tumors of NB rats and a theory of action. Cancer Res., 37:82-94, 1977.

47. Porter, E.H., Hewitt, H.B., and Blake, E.R. The transplantation kinetics of tumour cells. Brit. J. Cancer, 27:55-62, 1973.

48. Shellabarger, C.J. Induction of mammary neoplasia after in vitro exposure to x-rays. Proc. Soc. Exper. Biol. Med., 136:1103-1106, 1971.

49. Terzaghi-Howe, M. Inhibition of carcinogen-altered rat tracheal epithelial cell proliferation by normal epithelial cells in vivo. Carcinogenesis, 8:145-150, 1987.

50. Watanabe, H., Tanner, M.A., Domann, F.E., Gould, M.N., and Clifton, K.H. Inhibition of carcinoma formation and of vascular invasion in grafts of radiation-initiated thyroid clonogens by unirradiated thyroid cells. Carcinogenesis, 9:1329-1335, 1988.

51. Wolff, S., Afzal, V., Wiencke, J.K., Olivieri, G., and Michaeli, A. Human lymphocytes exposed to low doses of ionizing radiations become refractory to high doses of radiation as well as to chemical mutagens that induce double-strand breaks in DNA. Int. J. Radiat. Biol., 53:39-48, 1988.

THE SOMATIC MUTATIONAL COMPONENT OF HUMAN CARCINOGENESIS

Mortimer L. Mendelsohn

Somatic mutation is the prototypic mechanism for the immediate, irreversible, additive effects that are characteristic of initiation of carcinogenesis. Such genetic changes are often used, explicitly or implicitly, as the transition device between stages in quantitative multistage models designed to describe carcinogenesis. Recent developments in the genetics of cancer and in somatic mutagenesis greatly increase our understanding of these processes, and are worth reviewing in the context of such multistage modelling.

MODERN CONCEPTS OF THE GENE AND ITS MUTATION

A gene was initially understood as an abstraction representing the heritability and mutability of a specific biological characteristic. In this guise it was also viewed as the target for stochastic damaging agents. One could imagine a gene's relative size based on the likelihood of it being damaged from some overall background rate of mutation or from a dose of radiation or chemical mutagen. The detailed reality underlying these models has been provided by modern molecular biology, and with it has come several additional layers of insight and complication. A brief introduction to this subject is provided for the uninitiated reader and a good general text would be Molecular Biology of the Gene by Watson et al. [23].

The current view of a gene and its controls is diagrammed in Figure 1. The horizontal line represents one of the two complementary strands of a small stretch of DNA, the genetic molecule consisting of a string of four types of basepairs whose order is the genetic information. Genes are arranged discretely along the DNA, there being an estimated 100,000 genes in each haploid set of 3-billion basepairs in a diploid human cell. The essence of the gene is the coding region, the part that is definitively transcribed (copied) into messenger RNA and then translated into a corresponding string of amino acids making up the enzyme or structural protein that is the ultimate gene product.

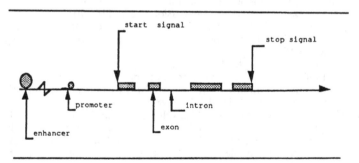

Figure 1. A typical gene.

The coding region begins at the left side of the leftmost box with a start signal and ends at the right side of the rightmost box with a stop signal. Each box is an exon, a region that is directly coded into protein. Between the boxes are non-coding regions called introns. These regions are spliced out of the RNA transcript, and are not translated into protein. he introns range from a handful of bases to tens of kilobases in length, and from zero to fifty or so in number. They are not present in prokaryotic cells such as bacteria, but are present universally in eukaryotes. Their function in higher organisms remains obscure, although they may provide a mechanism for easy exchange of exons among genes in the evolution of new or better gene products.

Upstream to the gene (to the left in the figure) is its promoter region. The promoter functions to turn the gene on and off via the competition between binding of various control elements and the polymerases that translate the coding region into messenger RNA. Further upstream, at a variable distance, are enhancer regions which modulate how vigorously the gene is translated.

The smallest mutation, a so called point mutation, is the change of a given base to any one of the three others. If this occurs in an exon, it will very likely cause a change in amino acid in the gene product. This may have a profound effect, no effect, or some gradient of effect in between. A profound effect occurs if, for example, the change blocks the active site of the enzymatic gene product. An intermediate effect might come from a change that slightly distorts the active region. No effect might happen if the change replaces one amino acid with a similar one in a non-sensitive scaffolding region of the protein. Base changes in the non-coding introns are likely to have no effect except in the rare circumstances when they generate a new stop signal (in which case the gene product is incomplete and likely to be non-functional) or when they obscure the information the cell uses to define the junction between intron and exon (in which case there is again likely to be loss of functional protein). Point mutations might also alter the promoter and enhancers, and conceivably disrupt their intervening regions.

Mutations can also arise from discontinuities in the DNA. These can be secondary to crossing within or between strands at regions of complementary sequences. Thus a polymerase might skip across a looped strand thereby deleting a stretch of DNA, or across homologous chromosomes making a region homozygous. DNA rearrangements also come from paired double-strand breaks which free a piece of DNA and allow it to invert or delete, or when the breaks are on separate chromosomes, to translocate. Discontinuities can have the same range of effects as point mutations, particularly when they are small, but the larger the distortion in the gene product, the more likely that it will be non-functional. Many discontinuities cause frame-shifts which disrupt triplet coding and lead frequently to non-functional gene product. Discontinuities may be large enough to affect several genes simultaneously. Rearrangements that are larger than genes can displace a gene from one place in the genome to another. When the controlling regions come with the gene, the relocation is likely to have no effect. However, separating a gene from its controllers or giving it access to new controllers can be a very significant event.

Mutagens are agents or substances known to change bases, break DNA strands or otherwise mutate genes. Chemical mutagens in general are activated *in vivo* to electrophilic form and then react with the DNA. They typically produce their mutational effect by causing miscoding during DNA replication, by causing error-prone repair, or by interfering with replication. Some chemical mutagens operate by being DNA analogues and are incorporated only during the DNA-synthetic part of the cell cycle. Repair mechanisms efficiently locate damaged DNA and return it to normal through a variety of mechanisms involving the opposite strand. Error-prone damage and repair generally occurs when the cell must make the choice between survival and mutation. Since errors are made permanent during DNA replication, there is a natural competition between repair and replication. The probability of mutation is heightened by the nearness in time of the damaging event to the cell's next attempt at DNA synthesis, thus providing yet another way for cell multiplication to modulate mutagenesis. Ionizing radiation produces most of its genetic damage through breakage of DNA strands, particularly double strand breaks leading to DNA rearrangements. Broken strands are repaired back to normal or misrepaired into DNA rearrangements rapidly and permanently. Gene function could be changed within minutes of exposure by this mechanism, however the bulk of genetic effect occurs later during the reassortment of chromosomes and fragments when the cell divides. Some chemicals are radiomimetic in their behavior, emphasizing rearrangements rather than base changes.

GENETIC CHANGES AND CANCER

Recent discoveries in the molecular genetics of cancer tell us that there are at least two types of genetic change involved in the process. See Knudsen's and Wiseman's articles in this volume for further detail.

Oncogenes are one of these types, and are products of mutational events that release a gene from down-control, i.e., they somehow increase the function of the gene. One way that this can happen is by point mutation. Increasing the function of a gene by directly changing its coding properties is an unlikely outcome of mutation, but this is precisely what happens in activating the ras family of proto-oncogenes by one of several very specific base changes. Another way a gene can be activated is by placing its coding region next to a powerful promoter. This happens in Burkitt's lymphoma, by a translocation which places the *myc* proto-oncogene under the control of a heavily turned on immunoglobulin promoter. A third and final example of this mechanism is gene amplification, an abnormal process in which the oncogene is copied over and over in the DNA and is made capable of great over-function. These events are all dominant in the sense that an over-functioning gene has its effect regardless of what its homologous partner is doing. A single activated oncogene does not appear to be enough to make a cell cancerous, and at least two different oncogenic changes are necessary. In the context of a two-stage model of carcinogenesis, hits in two different proto-oncogenes is a reasonable construct. However, while the number of known oncogenes continues to grow, there are still many cancers lacking recognizable oncogenes. Thus there is major uncertainty about the potential size of the combined target of all oncogenes, as well as about the fraction of cancers that are indeed initiated by this mechanism.

Tumor suppressor genes provide the second type of cancerous genetic change. Cells are normally homozygous (i.e., contain two copies) for some number of recessive cancer suppressors. When both copies of a suppressor are lost, the cell loses down-control of growth, and is likely to appear as one or another specific type of cancer. Either copy of the gene can be eliminated by any of the mechanisms that lead to loss of gene function. The disease retinoblastoma is the first and still the best example of this mechanism (see Knudson, this volume). Many cases of retinoblastoma stem from the inheritance of one inactive or missing copy of the retinoblastoma suppressor gene. Loss of the second copy by somatic mutation at an early age, when retinoblast cells are still active, is thought to cause this serious form of eye cancer. Such loss is sufficiently likely that it occurs in one eye in essentially all cases, and in both eyes in some 70% of cases [10]. Analysis of the tumor DNA reveals that both suppressor alleles are identically abnormal in something like 60% of tumors. This identity across homologous chromosomes applies as well to genes surrounding the suppressor and is probably due to a recombinational copying of the region containing the inherited mutation from the abnormal chromosome to its sister chromosome. Such somatic induction of local homozygosity is then another mechanism leading to expression of recessive mutation and of cancer suppressors.

Given this spectrum of mutational mechanisms for causing initiation, one would like to understand more of the genetic parameters that govern, describe or predict their behavior. Are human background mutation rates sufficient to explain background cancer rates via such stochastic and relatively simplistic mechanisms? Are there likely to be enough mutational events in the eyes of a child missing one retinoblastoma repressor gene to explain the loss of the other gene as well as the corresponding probability of retinoblastoma in one or both eyes? In the very few instances where dose-response information is available, are the mutational data consistent with the carcinogenesis data? Can one cross-calibrate from people with repair deficiency to those with chemical- or radiation-induced mutation? Or can one compare from one repair deficiency to another? Will somatic mutation provide any insight into species comparisons of carcinogenesis?

Such questions must eventually be answered by studying the particular proto-oncogenes and repressors directly for their mutagenicity and biological effects. However, because of the general predictability of mutational processes, reasonable answers can be obtained from any representational genes in the species in question. For the human, these would be those genes for which there are now useful assays for somatic mutation.

SOMATIC MUTATION ASSAYS

Four assay systems are now available to study somatic mutation in the human.

The HPRT assay is based on the gene *hprt* which codes for an enzyme, hypoxanthine phosphoribosyltransferase, in the precursor pathway of DNA synthesis [1,6,15]. The gene is on the X chromosome and has 9 exons and 34 kilobases from start to stop signal. Complete or near complete loss of function of this enzyme prevents the cell from phosphorylating thioguanine and succumbing to its toxic effects. Mutants can be either identified by autoradiographic detection of DNA synthesis or can be selectively cloned on the basis of their resistance to thioguanine. Once cloned, the cells can be tested for enzyme level, classified by lymphocyte type and analyzed to determine the underlying mutational defect in their DNA [16].

The HLA-A assay involves the gene *hla-a* which codes for one type of human lymphocyte antigen involved in self-recognition and transplantation rejection [8]. The gene is 5 kilobases long and is on chromosome 6. Antibodies specific for two common subtypes of *hla-a* can be used to detect or kill lymphocytes expressing the gene product on their cell surface. This permits the counting or cloning of rare cells that have lost gene expression. DNA lesions can be analyzed in the clonal version of the assay, and relatively simple tests for associated genes in the immune complex permit gauging the extent of genetic change in the clones.

The glycophorin A assay reflects damage to the corresponding gene on chromosome 4 [9]. The gene has 4 exons and is 44 kilobases long from start to stop signal. Glycophorin A is an abundant nonessential transmembrane glycoprotein of red blood cells. Human populations contain two equally frequent alleles in which the proteins differ by two amino acids near the extracellular terminus. With fluorescent antibodies that can discriminate the two forms of glycophorin A, the amount of each protein can be estimated in single red cells by flow cytometry. Heterozygotes (those people with one copy of each allele) are tested for frequency of cells that either have lost antibody recognition of one allele and have become effectively hemizygous or are expressing two copies of one allele and have become effectively homozygous. No cloning or DNA confirmation can be done in red blood cells.

The assay for mutant ß globin involves the 3 exon, 2 kilobase ß globin gene on chromosome 11 [17,20]. Many mutant variants of human ß globin, such as sickle hemoglobin, are well characterized as single amino-acid changes in the protein and single base changes in the DNA. Red blood cells which contain an allele's worth of specific mutant globin can be detected and counted by immunological and automated cytological methods. Again, cloning and DNA confirmation or not available.

Thus three of the four assays are based on complete or nearly complete loss of some aspect of gene function, analogous to loss of a repressor gene. In two of these, HPRT and HLA-A, the losses can be related back to damage in the DNA and generally involve either strategically located base changes or DNA rearrangements. In the HLA-A assay the rearrangements are frequently recombinational, again analogous to some cancer suppressors [22]. Roughly half of background and induced events detected using the glycophorin A assay are also recombinational. In the fourth assay, ß globin, the mutation is recognized as an altered gene product which by strong inference is secondary to a specific modification of a specific base in the DNA. This change is analogous to the similarly specific change in codon 12 or 60 of the *ras* oncogene.

BACKGROUND MUTATION RATES

Background rates are available for mutant frequencies at all four loci. In young to middle-aged normal subjects, for every million testable cells, there are approximately 30 cells with *hla-a* loss, 10 with *hprt* loss, 7 with glycophorin A loss and 0.03 expressing sickle hemoglobin. That the last value is as much as 1000 fold lower than the others is consistent with the mutational target for the ß globin assay being a single specific base. The variation among the other three background rates, however, does not correlate well with overall gene size since *hla-a* has the smallest size and the highest background rate. Other factors can also affect background frequency of mutants. The tests themselves have varying efficiency of detection, with HPRT being more stringent (i.e., detecting fewer mutants) than HLA-A. The tests also differ considerably in the spectrum of lesions they detect. Since *hprt* is on the X chromosome, it is in a functionally hemizygous (i.e., only one active copy) environment in which large deletions have a high likelihood of leading to cell lethality and recombinational events are either absent or unlikely. Also the loci may vary in how much of the gene is indifferent to mutational change. There is a real possibility that genes vary in their number and intensity of hot spots, a term for regions that are particularly sensitive to mutation. Finally, the frequency of mutant cells in a sample is a direct function of the persistence of the mutants over time. The data with the HPRT assay suggest decay of mutant frequencies within several months of any mutagenic exposure [2,14]. With glycophorin A mutants, radiation induction leads to persistence for decades, while chemically induced mutants disappear with the turnover time of red blood cells [4,5,7,11,13].

Closely related to background rate is the age structure of the background rate. Particularly with the long-lasting mutant cells, one would expect progressive accumulation of mutant cells as spontaneous mutations continue to appear over time. In fact, the HPRT and glycophorin A assays show a similar age structure of mutant frequency, increasing approximately 2% per year of age [6,9,20,21]. This low rate of increase probably reflects selection against *hprt* mutants, while for glycophorin A it is likely to be more a function of background having a large component of recent mutants coming from short-lived committed bone-marrow cells.

INDUCTION RATES

Radiation data are available for three of the somatic mutation tests. *In vitro* exposure of human peripheral blood lymphocytes induces 23 mutants per million cells per Gray with the HPRT assay [16], and 59 mutants per million cells per Gray with the HLA-A assay [8]. While these differ by a factor of two, the assays have essentially the same radiation doubling dose, roughly 0.7 Gray. Data *in vivo* from Japanese atomic bomb survivors is available for the HPRT [7] and glycophorin A assays [11,13], measured 40 to 44 years after the event. The HPRT results indicate 1 additional mutant per million cells per Gray, a rate that is 4% of the *in vitro* value and another indication of selection against the mutants. The glycophorin A results give roughly 40 additional mutants per million cells per Gray with a doubling dose of 0.25 Gray. The *in vitro* data for the two lymphocyte assays, and the *in vivo* data for glyocophorin A are in rough agreement with each other and with data from many other animal and plant genetic systems.

Chemotherapy data from cancer patients are also available for the HPRT [2,16] and glycophorin A [4,5] assays, but such studies lack the dosimetric detail of the radiation data. The best one can say is that the two assays are responding similarly with 4- to 8-fold increases in recently treated patients. Both assays show return of mutant frequencies to normal in several months, the HPRT presumably because of selection and the glycophorin A presumably because of effects predominantly in short-lived, actively cycling, committed bone marrow cells.

Effects from smoking are roughly similar with the HPRT and glycophorin A assays, with smokers showing 30-80% higher frequencies of mutant cells than nonsmokers [1,9,20].

OVERVIEW

In applying these data to models of carcinogenesis, it is important to understand that these somatic mutation assays are measuring mutant frequency and not mutation rates. The persistence problem is one aspect of this problem. Another is the finding that *hprt* mutants sometimes have identical cell types and DNA lesions, indicating that they have clonally amplified after sustaining the mutation [16]. It is not obvious which property the modeler really wants: mutation rate or mutant frequency. Mutation rate is a better descriptor of the underlying process of mutation production, but mutant frequency better describes the population at risk to subsequent compounding events.

With only four genes, some inconsistencies and the usual difficulties interpreting small populations of people, there is moderate uncertainty in any conclusions one might draw from these data. Nevertheless, there are useful generalizations for range finding and order of magnitude estimates of parameters.

A reasonable cumulative background level of human somatic mutant frequency for gene loss events (as in repressor loss) is 10^{-5} per gene. Similar or perhaps slightly lower values probably apply for recombinational events. Mutant cells accumulate slowly with age as a roughly linear function with a slope of 2% per year. For single base changes, as in point mutations of the *ras* oncogene, the corresponding background frequency is roughly 10^{-8}. Long-term survival of mutant cells can range from near 100% to 4%, depending on the gene, cell type, mutant type and inducer. Chemical induction is usually but not always related to cycling cells. Radiation induction is less sensitive to the cell cycle, and more likely to involve DNA rearrangements than is chemical induction. The yield of radiation induced mutants is approximately 40 per million cells per gene per Gray.

These values are variably consistent with cancer data. For retinoblastoma, the relative frequency of bilateral, unilateral and unaffected cases after inheritance of the retinoblastoma gene is consistent with the probability of 1 critical somatic mutant in the million or so retinoblast cells per eye [10]. This is comfortably within the 10^{-5} background frequency for either gene loss or recombination mutants that one would predict from somatic mutation assays. For the *ras*-like change, the 10^{-8} rate projects to around 20 point-mutated cells for each of the critical codons per mouse liver.

Another set of benchmarks comes from the growing body of data on human repair defectives. HPRT and glycophorin A assays indicate a 10-fold increase in frequency of mutant cells in homozygotes with ataxia telangiectasia [3,20]. These patients have multiple disorders, including a very high predilection for lymphoid cancers. The HPRT assay similarly shows a 10-fold increase in mutants in patients with xeroderma pigmentosum who have hypersensitivity to ultraviolet radiation effects and die of skin cancer if not protected. The glycophorin A assay gives normal results in this group, perhaps because the bone marrow is not exposed to ultraviolet light. In Bloom's syndrome, the glycophorin A assay scores 100-fold increases [12], and essentially all of these patients have cancer by middle age.

In contrast, atomic bomb survivors, with their roughly 2-fold increase in glycophorin A variants at 1 Gy, have an associated 4-fold increase in leukemia mortality and a 0.3-fold increase in mortality from all other cancers [19]. In a study done in San Francisco, randomly selected cancer patients prior to treatment do not have significantly increased frequencies of glycophorin A variants [5]. The statistical strength of this early result is sufficient to rule out average mutant frequencies of 1.25 times background or greater in such patients. This negative outcome and the quantitative inconsistencies or uncertainties in the repair defectives and A-bomb survivors raise important questions about the relationship between somatic mutation and cancer induction in the human. However, with the exception of the lack of increased mutants in cancer patients, the results are all supportive of some role of somatic mutation in carcinogenesis. Further studies should help to clarify these issues, as will the incorporation of realistic mutational data into the models for carcinogenesis.

30

Work performed under the auspices of the U.S. Department of Energy by the Lawrence Livermore National Laboratory under Contract W-7405-ENG-48.

1. Albertini, R.J., Sullivan, L.M., Berman, J.K., Greene, C.J., Stewart, J.A., Silveira, J.M. and O'Neill, J.P. Mutagenicity monitoring in humans by autoradiographic assay for mutant T lymphocytes. Mutat. Res. 204: 481-492, 1988.

2. Ammenheuser, M.M., Ward, J.B.Jr., Whorton, E.B.Jr., Killian, J.M. and Legator, M.S.. Elevated frequencies of 6-thioguanine-resistant lymphocytes in multiple sclerosis patients treated with cyclophosphamide: a prospective study. Mutat. Res., 204: 509-520, 1988.

3. Bigbee, W.L., Langlois, R.G., Swift, M. and Jensen, W.L.. Evidence for an elevated frequency of *in vivo* somatic cell mutations in ataxia telangiectasia. Am. J. Hum. Genet., 44: 402-408, 1989.
4. Bigbee, W.L., Langlois, R.G., Jensen, R.H., Wyrobek, A.W. and Everson, E.G.. Chemotherapy with mutagenic agents elevates the *in vivo* frequency of glycophorin A "null" variant erythrocytes. Environ. Mutagen. 9: 14, 1987.

5. Bigbee, W.L., Wyrobek, A.J., Langlois, R.G., Jensen, R.H. and Everson, R.B. The effect of chemotherapy on the *in vivo* frequency of glycophorin A "null" variant erythrocytes. Mutation Res. (in press, 1989).

6. Cole, J., Green, M.H.L., James, S.E., Henderson L. and Cole, H.. A further assessment of factors influencing measurements of thioguanine-resistant mutant frequency in circulating T-lymphocytes. Mutat. Res., 204: 493-507, 1988.

7. Hakoda, M., Akiyama, M., Kyoizumi, S., Awa, A.K., Yamakido, M.and Otake, M. Increased somatic cell mutant frequency in atomic bomb survivors. Mutation Res., 201: 39-48, 1988.

8. Janatipour, M., Trainor, K.J., Kutlaca, R., Bennett, G., Hay, J., Turner, D.R. and Morley, A.A. Mutations in human lymphocytes studied by an HLA selection system. Mutation Res. 198: 221-226, 1988.

9. Jensen, R.H., Bigbee, W.L. and Langlois, R.G.. *In vivo* somatic mutations in the glycophorin A locus of human erythroid cells. Banbury Report 28: Mammalian Cell Mutagenesis, 149-159, 1987.

10. Knudson, A.G., Jr. Mutation and cancer: statistical study of retinoblastoma. Proc. Nat. Acad. Sci.(USA), 68: 820-823, 1971.

11. Kyoizumi, S., Nakamura N., Hakoda, M, Awa, A.A., Bean, M.A., Jensen, R.H. and Akiyama, M. Detection of somatic mutations at the glycophorin A locus in erythrocytes of atomic bomb survivors using a single beam flow sorter. Cancer Res. 49: 581-588, 1989.

12. Langlois, R.G., Bigbee, W L., Jensen, R.H. and German J.L. Evidence for elevated *in vivo* mutation and somatic recombination in Bloom's syndrome. Proc. Nat. Acad. Sci. (USA) (In press, 1989).

13. Langlois, R.G., Bigbee, W.L., Kyoizumi, S., Nakamura, N., Bean, M.A., Akiyama, M. and Jensen, R.H.: Evidence for increased somatic cell mutations at the glycophorin A locus in atomic bomb survivors. Science 236: 445-448, 1987.

14. Mohrenweiser, H.W. and Jones, I.M. Review of the molecular characteristics of gene mutations of the germline and somatic cells of the human. Mutation Res. (In press, 1989.)

15. Morley, A.A., Trainor, K.J., Seshadri, R. and Ryall, R.B. Measurement of *in vivo* mutation in human lymphocytes. Nature (London), 302: 155-156, 1983.

16. Nicklas, J.A., O'Neill, J.P., Sullivan, L.M., Hunter, T.C., Allegretta, M. Chastenay, B.F., Libbus, B.L. and Albertini, R.J. Molecular analyses of *in vivo* hypoxanthine-guanine phosphoribosyltransferase mutations in human T-lymphocytes: II. Demonstration of a clonal amplification of *hprt* mutant T-lymphocytes *in vivo*. Envir. & Molec. Mutagen. 12, 271-284, 1988.

17. Papayannopoulou, Th., Mcguire, T.C., Lim, G., Garzel, E. and Stamatoyannopoulos, G. Identification of haemoglobin S in red cells and normoblasts, using fluorescent anti-Hb S antibodies. Brit J Haemat 34: 25-31, 1976.

18. Sanderson, B.J.S., Dempsey, J.L. and Morley, A.A. Mutations in human lymphocytes: Effect of X- and UV-irradiation. Mutat. Res., 140: 223-227, 1984.

19. Shimizu, Y., Kato, H., Schull, W.J., Preston, D.L., Fujita, S. and Price, D.A. Studies of the mortality of A-bomb survivors. 9. Mortality, 1950-1985: Part 1. Comparison of risk coefficients for site-specific cancer mortality based on the DS86 and T65DR shielded kerma and organ doses. Radiation Res., 118: 502-524. 1989.

20. Tates, A.D., Bernini, L.F., Natarajan, A.T., Ploem, J.S., Verwoerd, N.P., Cole, J., Green, M.H.L., Arlett, C.F. and Norris, P.N. Detection of somatic mutants in man: *hprt* mutations in lymphocytes and hemoglobin mutations in erythrocytes. Mutation Res., 213: 73-82, 1989.

21. Trainor, K.J., Wigmore, D.J., Chrysostomou, A., Dempsey, J.L., Seshadri, R. and Morley, A.A. Mutation frequency in human lymphocytes increases with age. Mech. Age Dev., 27: 83-86, 1984.

22. Turner, D.R., Grist, S.A., Janatipour, M. and Morley, A.A. Mutations in human lymphocytes commonly involve gene duplication and resemble those seen in cancer cells. Proc Nat Acad Sci (USA) 85: 3189-3192, 1988.

23. Watson, J.D., Hopkins, N.H., Roberts, J.W., Steitz, J. A. and Weiner A.M. Molecular Biology of the Gene, 4th Edition, The Benjamin/Cummings Publishing Co, Inc., Menlo Park, 1987.

Two-Event Carcinogenesis: Roles of Oncogenes and Antioncogenes
Alfred G. Knudson, Jr.

Abstract: A biological model for carcinogenesis that considers two genetic events as critical can be fitted well to age-specific incidence data on cancers of both children and adults. In the past few years advances in cancer genetics have disclosed two classes of genes that are the apparent targets of such events. For one class, the oncogenes, a single event can cause inappropriate activation, but both in vitro and in vivo data suggest that a second oncogene must also be activated. For the second class, the antioncogenes, two events are necessary for carcinogenesis, these involving inactivation or loss of both copies of a specific antioncogene. Thus, two events may be necessary for the induction of all forms of cancer. Alterations in other oncogenes and antioncogenes often occur in all cancers, but individually these may be neither necessary nor sufficient in carcinogenesis.

1. Introduction

The somatic mutation hypothesis for the origin of cancer can accommodate observations on environmental mutagens, including radiation, chemicals, and retroviruses. It also implies that there will always be a background incidence of cancer due to "spontaneous" mutations. More difficult to answer have been questions regarding the critical number of such genetic events, and the identity of the target genes that are mutated. The first of these questions has been the subject of epidemiologic models that have attempted to explain age-specific incidence data and of laboratory models that involve radiation, chemicals, and/or tumor viruses. The second of these questions has been answered at least in part by studies of transforming retroviruses and of hereditary cancer.

Epidemiologic models have been focused on age-specific incidence data for the common cancers of adults, such as carcinomas of colon and lung. The theme of these studies has been that the rise in incidence with the nth power or so of age could be explained by a succession of $n+1$ events, or by fewer than n events in conjunction with differential growth of cells that have sustained one or more event. The most parsimonious model, constructed by Moolgavkar and his colleagues (e.g., 45 and this volume), is of the latter type and calls for just two genetic events.

Two event models have also been favored by various laboratory investigators. Thus, the classical experiments on chemical carcinogenesis in rodent skin entailed two steps, initiation and promotion. The first event has been interpreted as a genetic event, whereas the second event has not. However, the end-point in these experiments was most often a papilloma. The yield of carcinomas is greatly increased if the promoter is followed by further application of initiator, suggesting that another event converts some cells to carcinomas. Both the experimental and mathemetical models have three features. Initiation and conversion may be considered as the two genetic events of the mathematical model, whereas promotion is a process that affects the growth of intermediate, once hit, cells.

In radiation experiments the long latent period has been interpreted as the time that elapses between an initiating event and a spontaneous, second mutational event. Experiments discussed at this meeting (Clifton, this volume) suggest that in vitro the first event is very common, and therefore cannot be what one usually considers to be a mutation. Perhaps such an effect is a gross chromosomal event, as occurs with genomic imprinting.

Transformation by acute transforming retroviruses in vitro occurs in a single step, but that phenomenon may be only partially relevant to the mechanisms of carcinogenesis that operate more generally. What I shall consider here is evidence for the nature and number of genetic events discovered in human cancers.

2. Oncogenes

Oncogenes were discovered as the genes necessary for transformation by acute oncogenic retroviruses. The subsequent discoveries of homologous genes in the cells of many animals, protooncogenes [59], and of cellular mechanisms for activating such genes, especially promoter insertion [21], enabled investigators of Burkitt's lymphoma to propose and validate a model of translocation-induced activation of a cellular protooncogene by juxtaposition to DNA sequences that normally activate immunoglobulin genes [19]. This model has guided studies of other translocations, notably those of chronic myelocytic leukemia, where a new, hybrid messenger RNA and protein product are synthesized [19].

The results obtained in these investigations suggest that carcinogenesis can occur in a single step, requiring simply a translocation between an oncogene-bearing chromosome and a chromosome carrying a tissue-specific activating sequence. However, it has been reported that a clonal abnormality has already occurred before the translocation in the case of chronic myelocytic leukemia. This involved the demonstration of chromosomal abnormalities, but not the specific translocation, in some lymphoid cells. Furthermore, these cells and the leukemic cells themselves expressed the same X-linked allele of glucose-6-phosphate dehydrogenase [13]. A tentative conclusion is that a first event in some lymphoid cell caused an abnormality in chromosomal behavior, while a second event involved the specific translocation. Perhaps such a situation also pre-dates the development of Burkitt's lymphoma. Preleukemic states of this disease are well known, but have not been well characterized.

Still it is apparent that these specific translocations must play a critical role in some cancers. Thus, in CML in its chronic phase, no other chromosomal aberrations are found in most cases. In the blastic phase other specific abnormalities occur and are clearly related to tumor progression. Specific translocations that are found early in the course of disease, and which are evidently important in generating a cancer, have been found in a number of leukemias, lymphomas, and sarcomas. For example, a translocation between chromosomes 11 and 22 is found in virtually every case of Ewing's sarcoma. It is interesting that such tumor-specific translocations have not been found in carcinomas. Are certain tissues better able to generate translocations? It is thought that translocations in lymphoid leukemias and lymphomas may depend upon the presence of certain enzymes, recombinases, that are active in normal lymphoid cells during the physiological recombination of immunoglobulin genes that makes specific antibody production possible. Do myeloid tissues and certain connective tissues also express such enzymes during their development, and do these tissues also rearrange some class of genes during normal development? Although carcinomas do not demonstrate such specific translocations, it is well known that carcinomas often have translocations of a wide variety. Perhaps the cancer cell loses the ability to suppress recombinase activity during the course of this progression.

The most frequent oncogene abnormality seen in cancer cells is a mutation in one of the family of ras genes. These mutations cluster at codon 12 and codon 60. In some cases the tumor or its metastases is mosaic for such a change, with some of the tumor cells not carrying it, and suggesting that the mutation was selected during tumor progression. In other circumstances, however, including some experimentally induced tumors in animals, these mutations are found at the earliest stages of the tumor and may be critical in generating the original clone of tumor cells. Not enough studies have been performed to conclude whether other specific genetic changes are required.

Ras mutations have been studied extensively in vitro. Transfection of DNA from tumor cells into NIH3T3 cells can transform them and permit the discovery of "activated" oncogenes in human tumors [55]. Analysis of such DNAs reveals that in most cases these are mutations in codon 12 or 60 of one of the ras genes. However, use of normal embryonic fibroblasts did not reveal transformation. Then it was found that co-transfection with an activated myc gene did cause transformation [36]. This finding has been interpreted to mean that NIH3T3 cells were unusual in that one step on the path to cancer had already been taken by them, whereas normal fibroblasts could be altered in an equivalent step by myc transfection. Transformation by oncogenes was therefore considered to involve two steps.

Another indication that oncogene-mediated carcinogenesis may involve two steps comes from the study of transgenic mice. Here an artificially activated oncogene is placed in the germline, and the animals are observed for tumors. A variety of tumors has been found. For example a myc gene construct with a mouse mammary tumor virus promoter leads to mammary tumors [60]. However, these tumors are clonal, indicating that at least one other, somatic, event is necessary for tumorigenesis.

Some other abnormalities of oncogenes commonly associated with human cancers are clearly involved in tumor progression. Chief among these is oncogene amplification, frequently involving one of the myc oncogenes. This amplification occurs at the DNA level and can lead to the presence of more than 100 copies of a gene per tumor cell.

3. Antioncogenes

The second kind of cancer gene, the antioncogene, was discovered through the study of hereditary cancer. At least one dominantly inherited form exists for most human cancers; the frequency of an inherited form may be as great as 50 per cent of all the cases of a relatively rare cancer or as small as one per cent or so of all cases of a common cancer. The inherited susceptibility is usually to one or a few types of cancer, not to cancer generally. Penetrance of the gene is usually incomplete, although this is often difficult to measure for cancers that typically occur at advanced age, where there is the confounding variable of competing mortality. The tumors that occur are not histologically unique to the hereditary form.

Incomplete penetrance and the small number of tumors in each carrier of predisposing germline mutations indicate that inheritance of such a mutation is not sufficient for oncogenesis. Question then arises regarding the number and nature of somatic events that are necessary to complete the process. Using retinoblastoma as a model tumor, Knudson proposed that a second, genetic event is necessary, and that the two events mutate or delete both copies of some gene [26,27]. This "two hit" model was subsequently applied to Wilms' tumor [30], neuroblastoma [29], and cancer generally [31]. The nonhereditary forms of these cancers were imagined to involve the same two events, both occurring somatically after conception.

Progress in testing this hypothesis came first for retinoblastoma and Wilms' tumor, because 3-5 per cent of cases of both of these tumors reveal a constitutional chromosomal deletion that points to the locus of the gene, 13q 14 for retinoblastoma [32] and 11p13 for Wilms' tumor [14]. The deletion in the former tumor also includes the gene for an enzyme, esterase D, which can be measured quantitatively in blood, and which exhibits a genetic polymorphism for an electrophoretic variant [57]. The variant permitted linkage studies in non-deletion cases that demonstrated homogeneity for the hereditary form; all cases involve the gene on 13q14 [58]. Evidence could also be elicited regarding the location of the second event. In some patients who were constitutionally heterozygous for the electrophoretic variant, only one allele was expressed in the tumor [17]. The conclusion was that one chromosome contained one esterase allele and a mutation limited to the retinoblastoma gene, whereas the other chromosome had been deleted for both the esterase gene and the retinoblastoma gene, in the tumor. Another case presented with only a 50 per cent level of esterase D in the tumor, probably due to a deletion in one chromosome, but the tumor had lost the other chromosome, so it contained no copy of the esterase D gene, or, presumably, the retinoblastoma gene [5].

Great progress ensued when Cavenee et al. [7] utilized restriction fragment length polymorphisms (RFLP'S) of DNA to gather indirect evidence on the nature of the second events. With several polymorphic loci on the chromosome in question, one can ascertain whether there is a difference between the constitutional genotype, as measured in blood cells, and the tumor. In more than half the cases one or more of the syntenic DNA loci has lost one allele, an expected outcome for large deletions, whole chromosomal loss, or somatic recombination. It was presumed that in the remaining cases a localized mutation had produced the second event, without loss of any syntenic DNA markers. As a result of this work the search for loss of heterozygosity in tumors became a tool for locating putative genes of this class.

Utilizing such probes for the examination of DNA extracted from tumors, Dryja [9] found that one probe could detect no DNA in some tumors, concluding that such tumors had sustained deletions of both copies of the gene he was probing. This would be the expectation for some tumors, if the probe were detecting the retinoblastoma gene itself, or a closely linked gene. Friend et al. [15] then used this probe to construct cDNA probes and found one tumor in which there was a deletion of both copies of an internal fragment of the gene. This and the finding that the cDNA probe detected messenger RNA in normal tissues but not in retinoblastomas led the investigators to conclude that this was indeed the gene itself. This work was quickly confirmed [16,38]. The retinoblastoma gene product is a protein of approximately 110 kd size that is found in the nucleus and binds to DNA [39]. The protein is either absent or abnormal in both retinoblastoma and osteosarcoma, a tumor that occurs in 10 per cent or so of hereditary cases of retinoblastoma. This discovery confirmed the mechanism of oncogenesis by mutation or loss of a gene. Because the normal copy of such a gene is in effect antioncogenic, the retinoblastoma gene has been referred to as an antioncogene [28], or tumor suppressor gene [25]. It is recessive in oncogenesis, although susceptibility to cancer is dominantly inherited; tumors are homozygously defective, whereas the host is heterozygous for the defect.

It appears that just these two events are necessary for production of a retinoblastoma. Although other chromosomal aberrations can be found in tumors, there is no indication that they are essential. Perhaps they improve the growth capability of the tumor. The tumor itself is derived from the embryonal retinoblast, a cell that normally differentiates during development of the retina. Once differentiation has occurred, no further mitoses are possible, which explains the incidence of the tumor in children. The "two hits" must occur before differentiation is completed.

Although the retinoblastoma gene is the only gene of this class that has been cloned, there is substantial evidence for the existence of other such genes. The early discovery of 11p13 deletion in some cases of Wilms' tumor and the generation of a series of RFLP's on chromosome 11 has made that tumor a subject of intense study. Loss of heterozygosity for syntenic RFLP's also occurs in about 60 per cent of cases of this tumor [12,34,48,49]. Complications have arisen because in some cases loss of heterozygosity occurs only in the distal region of 11p, at 11p15. This finding has been interpreted to mean that two genetically different forms of the tumor occur, one involving a gene at 11p13, the other a gene at 11p15. Furthermore, the study of families with multiple cases of Wilms' tumor has revealed a lack of linkage of the predisposing gene to either 11p13 or 11p15 [18,24]! This could be due to the existence of a third locus, but this conclusion is not the only one available. Another unexpected finding suggests an alternative explanation. This finding is that when loss of heterozygosity is observed in Wilms' tumors the maternal chromosomal segment is the one that is lost. Presumably the paternally derived chromosome selectively sustains an event that does not cause loss of chromosomal material, whereas the maternally derived one often does. This finding has led to the proposal that a different kind of event is more common for the paternal chromosome. This event is thought not to alter the nucleotide sequence, but to inactivate the gene by a process such as methylation. This preferential treatment of a chromosome derived from just one particular parent, in this case the father, has been referred to as imprinting. Presumably such imprinting can be caused by either genetic or environmental means. It has been proposed that the anomalous inheritance of Wilms' tumor may be due to a germline mutation that effects this imprinting of the paternally derived 11p gene [52]. It may be that such a process could also explain the very common "first event" observed in vitro during radiation carcinogenesis.

The pattern of discovery of candidate antioncogenes has been to find that the same chromosomal location is involved in some heritable and some non-heritable cases. In retinoblastoma and Wilms' tumor constitutional deletions are well known, and those same deletions are found in just the tumors in some nonhereditary cases. A similar situation exists for meningioma and acoustic neuroma. Deletion or monosomy of chromosome 22 is common, as is loss of heterozygosity for syntenic markers [44,53]. A constitutional deletion of chromosome 22 was found in one patient who died with multiple meningiomas [2]. Furthermore, central neurofibromatosis, which predisposes to both tumors, is attributable to a gene that has been mapped by constitutional abnormality and by genetic linkage to that same chromosome [10,50]. There is no direct evidence that both copies of the gene are mutant or lost in the tumors, but that is a reasonable explanation. There is no evidence that any specific further genetic changes are necessary for tumor formation.

A family has been reported with a high incidence of renal carcinomas and with mendelian transmission of a constitutional translocation that involved chromosome 3, with the break point at 3p14 [8,64]. In nonhereditary renal carcinomas deletion, translocation, and/or loss of heterozygosity for 3p RFLP's is an almost universal finding [35,67]. In addition, the dominantly heritable von Hippel-Lindau syndrome, which is strongly associated with renal carcinoma, often bilateral, has been mapped by linkage studies to chromosome 3p [54]. Here again there is a good case for a recessive tumor suppressor gene, or antioncogene, that has a narrow tissue specificity.

A common cancer, carcinoma of the colon, has also been the subject of considerable attention. In this disease there are at least two different dominantly heritable conditions that predispose to the tumor; one is familial polyposis coli (FPC), the other is familial non-polyposis colon carcinoma (FNPCC). The former is characterized by intermediate clonal lesions, adenomatous polyps, which regularly occur in the thousands in this disease. Carcinomas invariably occur by the age of 40 years in the absence of colectomy. It is not clear whether carcinomas occur only in polyps, or whether they can arise from the flat epithelium. In the second syndrome intermediate lesions are not characteristic, the carcinomas are most often in the more proximal half of the colon (in contrast to FPC, where most are in the distal colon), and endometrial carcinoma is often seen in female carriers of the mutation. It has been proposed that at least two nonhereditary forms of colon cancer occur, one initiated by mutation at the FPC locus and one, at the FNPCC locus [28].

One patient with FPC and other abnormalities was found to have a constitutional deletion in one chromosome arm 5q [22]. This observation led to a study of linkage and the discovery that indeed FPC is linked to some RFLP's in that chromosomal arm [6,41]. FNPCC gene has not yet been localized. Investigation of nonhereditary colon carcinomas has shown a 19-36 per cent loss of heterozygosity of RFLP's on 5q [56]. Since one expects that some losses of tumor suppressor genes are caused by local lesions that do not affect syntenic RFLP's, it is likely that 50 per cent or so of cases are initiated by mutation or loss at the FPC locus. Furthermore, some nonhereditary polyps should show such a loss too. Since the polyp is an intermediate lesion, it may result from just one mutation at the FPC locus, whereas the carcinomas would involve mutation or loss of both copies of the normal gene. When both mutations are local, there should be no loss of heterozygosity in either the polyp or the carcinoma. If the first is local and the second gross (as with deletion, monosomy, or recombination), only the carcinoma should show loss of heterozygosity for 5q RFLP's. If the first is gross and the second local, then both would show such loss. The incidence of loss in adenomas should therefore be lower than in carcinomas, which has been observed [62].

In colon cancers that begin with mutation at the FNPCC locus, whether germinal or somatic, there should of course be no loss of heterozygosity for 5q RFLP's. This idea cannot be tested directly, because no phenotypic differences have been found between carcinomas initiated at the two different loci. There is some indirect evidence, however. This relates to the use of myc oncogene expression in colon carcinomas. About 60-70 per cent of colon carcinomas have been found to express elevated levels of this oncogene. It has also been noted that the incidence of such expression is much lower in tumors of the proximal colon [3]. Astrin and her colleagues have therefore supposed that those carcinomas that express elevated levels of myc have arisen as the result of somatic mutation or loss of both copies of the FPC gene, and those that do not show such expression result from somatic mutation or loss of both copies of the FNPCC gene [3]. Accordingly, myc expressing tumors should show a significant incidence (50 per cent or so) of loss of heterozygosity for 5q

RFLP's, while tumors without elevated expression should not show such loss. This was indeed the case in a series of 27 tumors; of 19 with elevated myc expression, 9 showed 5q losses, whereas none of 8 without elevated myc showed such losses [3,11,62,63]. It may be relevant that in polyposis the cells of the colon synthesize DNA at every level of the crypts, indicating that even heterozygosity for the mutation has an effect, possibly mediated through increased myc expression.

The story of colon carcinoma has become more complicated, however. Genetic alterations have been found frequently at other loci, namely oncogenic mutations of K-ras and N-ras, and loss of heterozygosity of RFLP's on chromosomes 17p and 18q [4,37,47]. Such changes have been found in both polyps and in carcinomas. Loss of heterozygosity at 17p RFLP's occurs in about 80 per cent of colon carcinomas, so it appears to be important in all forms of colon carcinoma. These changes may provide a growth advantage to cells of both adenomas and carcinomas.

Chromosomal deletions and/or loss of heterozygosity of RFLP markers have been found frequently in other cancers, including carcinomas of breast and lung. Actual abnormality at the retinoblastoma locus has been found in some breast carcinomas [40,61] and some small cell carcinomas of the lung [20]. Loss of 11p markers has been found in some breast carcinomas [1,42]. Loss of 17p markers has also been found in breast cancers [43], suggesting that 17p may affect tumor growth in several cancers. Both deletions of 3p and loss of heterozygosity for 3p markers has been reported as common in all forms of lung cancer [33,46,66]. As with colon carcinomas, it seems that many of the adult carcinomas show losses that could be interpreted as losses of tumor suppressor genes. Which of these changes are essential for the various cancers is not known.

4. Conclusions

For some cancers, particularly those of hematopoietic and connective tissue origin, mutation or chromosomal rearrangement at an oncogene locus seems to be crucial, although more than one genetic change may be necessary. For several rare tumors of children and for some cancers of adults an antioncogene mechanism appears to operate. By its nature this kind of abnormality requires at least two events, those necessary to alter or delete the two copies of a suppressor locus. It may therefore prove to be the case that two events may be necessary for all forms of cancer, at least of the nonhereditary form (hereditary forms may require only one somatic event).

The multiplicity of genetic changes that are observed, especially in the carcinomas of adults, complicates interpretation. One interpretation is that these tumors are formed as a result of several (more than two) necessary events. Another interpretation is that just two events, those affecting some critical antioncogene, are necessary, and that the other events provide growth advantage to intermediate, once hit cells, or to the cancer cells themselves, i.e. contribute to tumor progression. Discrimination between these interpretations may be difficult. One kind of approach concerns reversion of tumorigenicity in vitro. It has been shown, for example, that fusion of a Wilms' tumor cell with a minicell containing chromosome arm 11p caused such reversion [65], as did infection of a retinoblastoma cell with a retroviral construct that included a copy of normal retinoblastoma cDNA [23]. These experiments indicate that loss of genes at these sites is critical for oncogenicity. One would presume that supplying a normal copy of the FPC gene to a colon carcinoma associated with mutation or loss of both of these copies should accomplish a similar result. One could further suppose that if supplying another missing gene, such as one at 17p, also caused reversion, then both loci were critical. On the other hand, if the latter maneuver did not have an effect, then one might conclude that only the 5q mutation was critical.

The fact that cancers of adults are more complicated than are the hematopoietic or pediatric malignancies could relate to the kinetic state of the target tissues. For example, in retinoblastoma the tumor arises from a target cell that rapidly reproduces during early development. A first somatic mutation in such a cell leads to mutant progeny cells, which can be quite numerous, thereby providing more targets for a second event. In adult renewal tissues, such as colon epithelium, the basal stem cell number does not normally change, so a mutation causes just one mutant cell. However, if an environmental stimulus, or a particular genetic change, such as a ras mutation, were to cause an increase in the number of once-hit stem cells, then, again, more cells would be available for a second event. Yet such growth stimulation of intermediate cells could be viewed as non-essential, even though it could greatly increase the probability of cancer.

5. Acknowledgements

This work was supported by NIH grants CA-06927 and CA-43211 and by an appropriation from the Commonwealth of Pennsylvania.

6. References

1. Ali, I.U., Lidereau, R., Theillet, C., Callahan, R. (1987). Reduction to homozygosity of genes on chromosome 11 in human breast neoplasia. Science, 238:185-188.

2. Arinami, T., Kondo, I., Hamaguchi, H., Nakajima, S. (1986). Multifocal meningiomas in a patient with a constitutional ring chromosome 22. J. Med. Genet., 23:178-180.

3. Astrin, S.M., Costanzi, C. (1989). The molecular genetics of colon cancer. Sem. Oncol., 16:138-147.

4. Baker, S.J., Fearon, E.R., Nigro, J.M., Hamilton, S.R., Preisinger, A.C., Jessup, J.M., van Tuinen, P., Ledbetter, D.H., Barker, D.F., Nakamura, Y., White, R., Vogelstein, B. (1989). Chromosome 17 deletions and p53 gene mutations in colorectal carcinomas. Science, 244:217-221.

5. Benedict, W.F., Murphree, A.L., Banerjee, A., Spina, C.A., Sparkes, M.D., Sparkes, R.S. (1983). Patient with 13 chromosome deletion: evidence that the retinoblastoma gene is a recessive cancer gene. Science, 219:973-975.

6. Bodmer, W.F., Bailey, C.J., Bodmer, J., Bussey, H.J.R., Ellis, A., Gorman, P., Lucibello, F.C., Murday, V.A., Rider, S.H., Scambler, P., Sheer, D., Solomon, E., Spurr, N.K. (1987). Localization of the gene for familial adenomatous polyposis on chromosome 5. Nature, 328:614-616.

7. Cavenee, W.K., Dryja, T.P., Phillips, R.A., Benedict, W.F., Godbout, R., Gallie, B.L., Murphree, A.L., Strong, L.C., White, R.L. (1983). Expression of recessive alleles by chromosomal mechanisms in retinoblastoma. Nature, 305:779-784.

8. Cohen, A.J., Li, F.P., Berg, S., Marchetto, D.J., Tsai, S., Jacobs, S.C., Brown, R.S. (1979). Hereditary renal-cell carcinomas associated with a chromosomal translocation. N. Eng. J. Med., 301:592-595.

9. Dryja, T.P., Rapaport, J.M., Joyce, J.M., Petersen, R.A. (1986). Molecular detection of deletions involving band q14 of chromosome 13 in retinoblastomas. Proc. Natl. Acad. Sci. USA, 83:7391-7394.

10. Duncan, A.M., Partington, M.W., Soudek, D. (1987). Neurofibromatosis in a man with a ring 22: in situ hybridization studies. Cancer Genet. Cytogenet., 25:169-174.

11. Erisman, M.D., Scott, J.K., Astrin, S.M. (1989). Evidence that the familial adenomatous polyposis is involved in a subset of colon cancers with a complementable defect in c-myc regulation. Proc. Natl. Acad. Sci. USA, 86:4264-4268.

12. Fearon, E.R., Vogelstein, B., Feinberg, A.P. (1984). Somatic deletion and duplication of genes on chromosome 11 in Wilms' tumours. Nature, 309:176-178.

13. Fialkow, P.J., Martin, P.J., Najfeld, V., Penfold, G.K., Jacobson, R.J., Hansen, J.A. (1981). Evidence for a multistep pathogenesis of chronic myelogenous leukemia. Blood, 58:158-163.

14. Francke, U., Holmes, L.B., Atkins, L., Riccardi, V.M. (1979). Aniridia-Wilms' tumor association: evidence for specific deletion of 11p13. Cytogenet. Cell Genet., 24:185-192.

15. Friend, S.H., Bernards, R., Rogelj, S., Weinberg, R.A., Rapaport, J.M., Albert, D.M., Dryja, T.P. (1986). A human DNA segment with properties of the gene that predisposes to retinoblastoma and osteosarcoma. Nature, 323:643-646.

16. Fung, Y-KT., Murphree, A.L., T'Ang, A., Qian, J., Hinrichs, S.H., Benedict, W.F. (1987). Structural evidence for the authenticity of the human retinoblastoma gene. Science, 236:1657-1661.

17. Godbout, R., Dryja, T.P., Squire, J., Gallie, B.L., Phillips, R.A. (1983). Somatic inactivation of genes on chromosome 13 is a common event in retinoblastoma. Nature, 304:451-453.

18. Grundy, P., Koufos, A., Morgan, K., Li, F.P., Meadows, A.T., Cavenee, W.K. (1988). Familial predisposition to Wilms' tumour does not map to the short arm of chromosome 11. Nature, 336:374-376.

19. Haluska, F.G., Tsujimoto, Y., Croce, C.M. (1987). Oncogene activation by chromosome translocation in human malignancy. Ann. Rev. Genet., 21:321-345.

20. Harbour, J.W., Lai, S-L., Whang-Peng, J., Gazdar, A.F., Minna, J.D., Kaye, F.J. (1988). Abnormalities in structure and expression of the human retinoblastoma gene in SCLC. Science, 241:353-357

21. Hayward, W.S., Neel, B.G., Astrin, S.M. (1981). Activation of a cellular onc gene by promoter insertion in ALV-induced lymphoid leukosis. Nature, 290:475-480.

22. Herrera, L., Kakati, S., Gibas, L., Pietrzak, E., Sandberg, A.A. (1986). Brief clinical report: Gardner syndrome in a man with an interstitial deletion of 5q. Am. J. Hum. Genet., 25:473-476.

23. Huang, H-J.S., Yee, J.-K., Shew, J.-Y., Chen, P.-L., Bookstein, R., Freidmann, T., Lee, E. Y.-H.P., Lee, W.-H. (1988). Suppression of the neoplastic phenotype by replacement of the RB gene in human cancer cells. Science, 242:1563-1566.

24. Huff, V., Compton, D.A., Chao, L.-Y., Strong, L.C., Geiser, C.F., Saunders, G.F. (1988). Lack of linkage of familial Wilms' tumour to chromosomal band 11p13. Nature, 336:377-378.

25. Klein, G. (1987). The approaching era of the tumor suppressor genes. Science 238:1539-1545.

26. Knudson, A.G. (1971). Mutation and cancer: statistical study of retinoblastoma. Proc. Natl. Acad. Sci. USA, 68:820-823.

27. Knudson, A.G. (1978). Retinoblastoma: a prototypic hereditary neoplasm. Semin. Oncol., 5:57-60.

28. Knudson, A.G. (1985). Hereditary cancer, oncogenes, and antioncogenes. Cancer Res., 45:1437-1443.

29. Knudson, A.G., Strong, L.C. (1972). Mutation and cancer: neuroblastoma and pheochromocytoma. Am. J. Hum. Genet., 24:514-532.

30. Knudson, A.G., Strong, L.C. (1972). Mutation and cancer: a model for Wilms' tumor of the kidney. J. Natl. Cancer Inst., 48:313-324.

31. Knudson, A.G., Strong, L.C., Anderson, D.E. (1973). Heredity and cancer in man. Prog. Med. Genet., 9:113-158.

32. Knudson, A.G., Meadows, A.T., Nichols, W.W., Hill, R. (1976). Chromosomal deletion and retinoblastoma. N. Engl. J. Med., 295:1120-1123.

33. Kok, K., Osinga, J., Carritt, B., Davis, M.B., van der Hout, A.H., van der Veen, A.Y., Landsvater, R.M., de Leij, L.F.M.H., Berendsen, H.H., Postmus, P.E., Poppema, S., Buys, C.H.C.M. (1987). Deletion of a DNA sequence at the chromosomal region 3p21 in all major types of lung cancer. Nature, 330:578-581.

34. Koufos, A., Hansen, M.F., Lampkin, D.B., Workman, M.L., Copeland, N.G., Jenkins, N.A., Cavenee, W.K. (1984). Loss of alleles at loci on human chromosome 11 during genesis of Wilms' tumour. Nature, 309:170-172.

35. Kovacs, G., Erlandsson, R., Boldog, F., Ingvarson, S., Müller-Brechlin, R., Klein, G., Sümegi, J. (1988). Consistent chromosome 3p deletion and loss of heterozygosity in renal cell carcinoma. Proc. Natl. Acad. Sci. USA, 85:1571-1575.

36. Land, H., Parada, L.F., Weinberg, R.A. (1983). Cellular oncogenes and multistep carcinogenesis. Science, 222:771-778.

37. Law, D.J., Olschwang, S., Monpezat, J-P., Lefrancois, D., Jagelman, D., Petrelli, N.J., Thomas, G., Feinberg, A.P. (1988). Concerted nonsyntenic allelic loss in human colorectal carcinoma. Science, 241:961-965.

38. Lee, W-H., Bookstein, R., Hong, F., Young, L-J., Shew, J-Y., Lee, EY-HP. (1987). Human retinoblastoma susceptibility gene: cloning, identification, and sequence. Science, 235:1394-1399.

39. Lee W-H., Shew, J-Y., Hong, F.D., Sery, T.W., Donoso, L.A., Young, L-J, Bookstein, R., Lee, EY-HP. (1987). The retinoblastoma susceptibility gene encodes a nuclear phosphoprotein associated with DNA binding activity. Nature, 329:642-645.

40. Lee, E.Y-H.P., To, H., Shew, J-Y., Bookstein, R., Scully, P., Lee, W-H. (1988). Inactivation of the retinoblastoma susceptibility gene in human breast cancers. Science, 241:218-221.

41. Leppert, M., Dobbs, M., Scambler, P., O'Connell, P., Nakamura, Y., Stauffer, D., Woodward, S., Burt, R., Hughes, J., Gardner, E, Lathrop, M., Wasmuth, J., Lalouel, J-M., White, R. (1987). The gene for familial polyposis coli maps to the long arm of chromosome 5. Science, 238:1411-1415.

42. Lundberg, C., Skoog, L., Cavenee, W.K., Nordenskjöld, M. (1987). Loss of heterozygosity in human ductal breast tumors indicates a recessive mutation on chromosome 13. Proc. Natl. Acad. Sci. USA, 84:2372-2376.

43. Mackay, J., Steel, C.M., Elder, P.A., Forrest, A.P.M., Evans, H.J. (1988). Allele loss on short arm of chromosome 17 in breast cancers. Lancet, 2:1384-1385.

44. Meese, E., Blin, N. Zang, K.D. (1987). Loss of heterozygosity and the origin of meningioma. Hum. Genet., 77:349-351.

45. Moolgavkar, S. H., Knudson, A. G. (1981). Mutation and cancer: a model for human carcinogenesis. J. Natl. Cancer Inst., 66:1037-1052.

46. Naylor, S.L., Johnson, B.E., Minna, J.D., Sakaguchi, A.Y. (1987). Loss of heterozygosity of chromosome 3p markers in small-cell lung cancer. Nature, 329:451-454.

47. Okamoto, M., Sasaki, M., Sugio, K., Sato, C., Iwama, T., Ikeuchi, T., Tonomura, A., Sasazuki, T., Miyaki, M. (1988). Loss of constitutional heterozygosity in colon carcinoma from patients with familial polyposis coli. Nature, 331:273-277.

48. Orkin, S.H., Goldman, D.S., Sallan, S.E. (1984). Development of homozygosity for chromosome 11p markers in Wilms' tumor. Nature, 309:172-174.

49. Reeve, A.E., Housiaux, P.J., Gardner, R.J.M., Chewings, W.E., Grindley, R.M., Millow, L.J. (1984). Loss of Harvey ras allele in sporadic Wilms' tumour. Nature, 309:174-176.

50. Rouleau, G.A., Wertelecki W, Haines, J.L., Hobbs, W.J., Trofatter, J.A., Seizinger, B.R., Martuza, R.L., Superneau, D.W., Conneally, P.M., Gusella, J.F. (1987). Genetic linkage of bilateral acoustic neurofibromatosis to a DNA marker on chromosome 22. Nature, 329:246-248.

51. Schroeder, W.T., Chao, L.-Y., Dao, D.D., Strong, L.C., Pathak, S., Riccardi, V., Lewis, W.H., Saunders, G.F. (1987). Nonrandom loss of maternal chromosome 11 alleles in Wilms' tumors. Am. J. Hum. Genet., 40:413-420.

52. Scrable, H., Cavenee, W., Ghavimi, F., Lovell, M., Morgan, K., Sapienza, C. (1989). A model for embryonal rhabdomyosarcoma tumorigenesis which involves genome imprinting. Proc. Natl. Acad. Sci. USA, in press.

53. Seizenger, B.R., Rouleau, G., Ozelius, L.G., Lane, A.H., St George-Hyslop, P., Huson, S., Gusella, J.F., Martuza, R.L. (1987). Common pathogenetic mechanism for three tumor types in bilateral acoustic neurofibromatosis. Science, 236:317-319.

54. Seizinger, B.R., Rouleau, G.A., Ozelius, L.J., Lane, A.H., Farmer, G.E., Lamiell, J.M., et al. (1988). Von Hippel-Lindau disease maps to the region of chromosome 3 associated with renal cell carcinoma. Nature, 332:268-269.

55. Shih, C., Padhy, L.C., Murray, M., Weinberg, R.A. (1981). Transforming genes of carcinomas and neuroblastomas introduced into mouse fibroblasts. Nature, 290:261-264.

56. Solomon, E., Voss, R., Hall, V., Bodmer, W.F., Jass, J.R., Jeffreys, A.J., Lucibello, F.C., Patel, I., Rider, S.H. (1987). Chromosome 5 allele loss in human colorectal carcinomas. Nature, 328:616-619.

57. Sparkes, R.S., Sparkes, M.C., Wilson, M.G., Towner, J.W., Benedict, W., Murphree, A.L., Yunis, J.J. (1980). Regional assignment of genes for human esterase D and retinoblastoma to chromosome band 13q14. Science, 208:1042-1044.

58. Sparkes, R.S., Murphree, A.L., Lingua, R.W., Sparkes, M.C., Field, L.L., Funderburk, S.J., Benedict, W.F. (1983). Gene for hereditary retinoblastoma assigned to human chromosome 13 by linkage to esterase D. Science, 219:971-973.

59. Stehelin, D., Varmus, H.E., Bishop, J.M., Vogt, P.K. (1976). DNA related to the transforming gene(s) of avian sarcoma viruses is present in normal avian DNA. Nature, 260:170-173.

60. Stewart, T.A., Pattengale, P.K., Leder, P. (1984). Spontaneous mammary adenocarcinomas in transgenic mice that carry and express MRV/myc fusion genes. Cell, 38:627-637.

61. T'Ang, A., Varley, J.M., Chakraborty, S., Murphree, A.L., Fung, Y.-K.T. (1988). Structural rearrangment of the retinoblastoma gene in human breast carcinoma. Science, 242:263-266.

62. Vogelstein, B., Fearon, E.R., Hamilton, S.R., Kern, S.E. Preisinger, A.C., Leppert, M., Nakamura, Y. White, R., Smits, A.M.M., Bos, J.L. (1988). Genetic alterations during colorectal tumor development. N. Engl. J. Med., 319:525-532.

63. Vogelstein, B., Fearon, E.R., Kern, S.E., Hamilton, S.R., Preisinger, A.C., Nakamura, Y., White, R. (1989). Allelotype of colorectal carcinomas. Science, 244:207-211.

64. Wang, N., Perkins, K.L. (1984). Involvement of band 3p14 in t(3:8) hereditary renal carcinoma. Cancer Genet. Cytogenet., 11:479-481.

65. Weissman, B.E., Saxon, P.J., Pasquale, S.R., Jones, G.R., Geiser, A.G., Stanbridge, E.J. (1987). Introduction of a normal human chromosome 11

into a Wilms' tumor cell line controls its tumorigenic expression. Science, 236:175-180.

66. Yokota, J., Wada, M., Shimosato, Y., Terada, M., Sugimura, T. (1987). Loss of heterozygosity on chromosomes 3, 13, and 17 in small-cell carcinoma and on chromosome 3 in adenocarcinoma of the lung. Proc. Natl. Acad. Sci. USA, **84**:9252-9256.

67. Zbar, B., Brauch, H., Talmadge, C., Linehan, M. (1987). Loss of alleles of loci on the short arm of chromosome 3 in renal cell carcinoma. Nature, 327:723-726.

Genetic Alterations During Carcinogenesis in Rodents:
Implications for Cancer Risk Assessment
Roger W. Wiseman

Abstract: Recent advances in the molecular genetics of
cancer in humans and experimental rodents provide strong
support for the somatic mutation theory of carcinogenesis.
It is now known that at least two distinct classes of
genes, proto-oncogenes and tumor suppressor genes, are
altered in cancer cells. A number of tumor types
frequently contain ras proto-oncogenes that have been
activated by a point mutation. Losses of heterozygosity
in specific chromosomal regions which are detected by the
analysis of restriction fragment length polymorphisms
suggest the involvement of tumor suppressor genes in a wide
variety of human cancers. This review discusses results
from a series of studies that have characterized activating
ras mutations in spontaneous and chemically induced liver
and lung tumors of mice and rats. Initial results from the
application of restriction fragment length polymorphism
analysis for tumor suppressor gene identification in liver
and lung tumors of B6C3F1 mice are also presented.

1. Introduction

The concept that carcinogenesis requires the
accumulation of multiple heritable alterations in tumor
cells is well established, and at least two distinct
classes of genes, proto-oncogenes and tumor suppressor
genes, have been implicated in the neoplastic process
[5,6,48]. Proto-oncogenes are normal cellular genes that
are activated in tumors and function as positive
proliferative signals for neoplastic cells. The hypothesis
that proto-oncogenes are involved in growth control of
normal and neoplastic cells is supported by increasing
evidence for links between oncogene products and growth
factors, growth factor receptors, and signal transduction
pathways [4,6,48]. On the other hand, tumor suppressor
genes are normal cellular genes that appear to act as
negative regulators of tumor cell proliferation and must
be lost, inactivated, or mutated in neoplastic cells. The
role of tumor suppressor genes in carcinogenesis is

supported by several lines of evidence [11,25,34], which
include suppression of tumorigenicity in cell hybrids,
studies of genetic predisposition to cancer in humans and
animals, and nonrandom chromosome deletions or losses of
heterozygosity for specific chromosomal regions (see
Knudson, this volume).

One of the most common genetic abnormalities that has
been detected in human and experimental rodent tumors is
the activation of ras proto-oncogenes by point mutations at
one of several sites within their coding regions (reviewed
in Refs. 3, 4, and 7). Considerable insights into the
structural and biochemical properties of the proteins
encoded by the ras proto-oncogene family have been
elucidated, but their basic function, which is believed to
involve signal transduction, has remained elusive [4]. Two
members of this family, H-ras and K-ras, were originally
identified as the normal cellular homologues of the viral
oncogenes of the Harvey and Kirsten murine sarcoma viruses,
respectively [4]. The third member, N-ras, was isolated
from a human neuroblastoma based on its transforming
activity in the NIH/3T3 transfection assay, which appears
to be especially sensitive to mutated ras genes [4]. In
this technique high molecular weight tumor DNAs are
introduced as a calcium phosphate co-precipitate into the
NIH/3T3 cell line and assayed for the ability to induce
morphological transformation or tumorigenicity in nude
mice.

Although ras activation has only been observed in 10-
15% of human cancers [7], it is extremely common in several
major tumor types. For example, activated K-ras genes have
been detected in more than 80% of pancreatic carcinomas
[1,7] and in about one third of lung adenocarcinomas [39].
Likewise, two groups [8,17] have found that 40-50% of human
colorectal tumors contain activated K-ras genes. The
results from a number of studies that have applied the
NIH/3T3 assay to experimental rodent systems are summarized
in Table 1. This brief survey shows that ras activation
has been observed in a variety of tissues and in many
different carcinogenesis models, ranging from in vitro cell

Table 1. Summary of proto-oncogene activation in experimental carcinogenesis.

Model	Target Tissue	Carcinogen	Proto-oncogene	Ref.
In Vitro Cell	C3H10T1/2	Methylcholanthrene	K-ras	32
Transformation	Guinea Pig	Several	N-ras	12
	Syrian Hamster	Several	H-ras, Unk.[a]	20
Spontaneous	Mouse Liver	None	H-ras	37
	Mouse Lung	None	K-ras	53
Single Dose	Rat Mammary Gland	Methylnitrosourea	H-ras	54
		7,12-Dimethylbenz-[a]anthracene	H-ras	54
	Mouse Liver	N-Hydroxy-2-acetyl-aminofluorene	H-ras	50
		Vinyl carbamate	H-ras	50
	Mouse Lung	Methylnitrosourea	K-ras	53
		Ethyl carbamate	K-ras	53
Multiple Dose	Mouse T-lymphoma	Methylnitrosourea	N- & K-ras	29
		Gamma-irradiation	K- & N-ras, Unk.	29
Initiation - Promotion	Mouse Skin	7,12-Dimethylbenz-[a]anthracene and 12-O-tetradecanoyl-phorbol-13-acetate	H-ras	35
Chronic Admin-	Rat & Mouse Lung	Tetranitromethane	K-ras	40
istration	Mouse Liver	Furan	H- & K-ras,	37
		Furfural	H- & K-ras, Unk.	37
Transplacental	Rat Schwanoma	Ethylnitrosourea	neu	33

[a]Unk.: Unknown

transformation to animal studies with chronic carcinogen administration.

Detection of the loss of heterozygosity in specific chromosomal regions by analyzing restriction fragment length polymorphisms (RFLP) has implicated tumor suppressor genes in a wide range of human neoplasia [11,34]. These reductions to homozygosity are believed to reflect the loss of the second normal tumor suppressor gene allele by chromosomal mechanisms (deletion, nondisjunction, or somatic recombination) after one allele has been inactivated by a deletion or point mutation [10]. This hypothesis has been confirmed in a series of studies on human retinoblastomas [11,48] and in work by Vogelstein

and his colleagues [45] on sporadic human colorectal
cancer. Neoplastic progression in the colon involves a
series of clearly defined phenotypic stages. K-ras
activation appears usually to occur in early adenomas and
it has been suggested that these lesions generally precede
the development of malignancy [8,17,45]. In addition,
about half of the advanced adenomas and three-quarters of
the carcinomas examined by these workers [45] have lost one
copy of a specific region of chromosome 18; a similar
number of carcinomas have lost one copy of chromosome 17p
sequences that include the p53 gene. Recently this group
demonstrated that the remaining p53 allele in colon tumors
that have lost one copy of chromosome 17p contains point
mutations believed to inactivate its normal function [2].
These observations, together with recent studies indicating
that wild type p53 protein has tumor suppressor activity
[16], support the concept that losses of heterozygosity are
indicative of the loss of genes that normally suppress
tumorigenesis.

This review will focus on genetic alterations in liver
and lung tumors of mice and rats. The importance of these
target organs for risk assessment is highlighted by a
recent summary of 350 long-term bioassays of chemicals by
the National Toxicology Program [23]. This study shows
that positive carcinogenic responses in the B6C3F1 mouse
are most frequently detected in the liver and lung. First,
I will discuss ras activation in these tumors. Then an
approach will be presented for the study of tumor
suppressor gene involvement in carcinogenesis of B6C3F1
mice.

2. Activation of ras Proto-oncogenes

Reynolds and coworkers [36] observed that spontaneous
liver tumors of B6C3F1 mice, which develop with an
incidence of about 30% at two years of age, frequently
contain transforming activity in the NIH/3T3 assay (Table
2). In contrast, DNA from a variety of spontaneous tumors
of F344 rats (not including liver tumors) were inactive in
this transformation assay. Southern analysis of the
NIH/3T3 transformants derived from these mouse tumors

demonstrated that both hepatocellular adenomas and carcinomas contained activated H-ras proto-oncogenes. Since normal liver DNA lacks transforming activity, a germ line ras mutation in the B6C3F1 mouse was ruled out.

These results were intriguing since previous studies [54] of rat mammary carcinomas induced by N-methyl-N-nitrosourea or 7,12-dimethylbenzanthracene had suggested that direct mutagenic effects of the carcinogens were responsible for H-ras activation. The initial observations with spontaneous B6C3F1 liver tumors were quickly extended to chemically induced tumors from the neonatal liver model by Wiseman et al. [50]. This system was especially attractive because administration of a single dose of a wide variety of chemical carcinogens to 12-day-old male B6C3F1 mice results in the formation of multiple liver tumors by 7-9 months in the absence of any further treatment [46,50]. The goal of this study was to determine whether the ras mutations result from the direct interaction of chemical carcinogens or from spontaneous errors in DNA replication during the carcinogenic process. We reasoned that structurally diverse carcinogens would yield distinct patterns of base substitution if the activating mutations resulted from the direct interaction of their electrophilic derivatives with the H-ras gene. On the other hand, if these alterations resulted from spontaneous mutagenic events, the distribution of base substitutions would be similar regardless of the chemical treatment.

We examined a series of B6C3F1 liver tumors induced by a single dose of the following carcinogens: N-hydroxy-2-acetylaminofluorene, vinyl carbamate, 1'-hydroxy(2',3') dehydroestragole, and diethylnitrosamine [41,50]. Each of 25 tumor DNAs induced by the first three carcinogens were able to induce morphological transformation when transfected into NIH/3T3 cells while less than half of diethylnitrosamine-induced tumors (14/32) exhibited transforming activity (Table 2). Southern analysis of these NIH/3T3 transformants demonstrated that all but one contained activated H-ras proto-oncogenes. The activating

Table 2. Activating mutations observed in the H-ras proto-oncogene of B6C3F1 mouse liver tumors.[a]

Treatment	Freq.[b]	Codon 61 (CAA)			Codon 117 (AAG)		Codon 13 (GGC)		Other
		AAA	CTA	CGA	AAC	AAT	CGC	GTC	
N-Hydroxy-2-acetyl-aminofluorene	7/7	7							
Vinyl Carbamate	7/7		6	1					
1'-Hydroxy(2',3') dehydroestragole	11/11		5	5					1 K-ras
Diethylnitrosamine	14/32	7	4	3					
Furfural	13/16	5		1		1	1	1	1 K-ras 3 Unk.[c]
Furan	13/29	4		1	2	2			2 K-ras 1 H-ras
Spontaneous	16/27	9	3	3					1 Unk.

aData taken from References 37, 41, and 50. The first four compounds were given as a single intraperitoneal injection to 14-day-old male mice. The remaining tumors were obtained from two-year chronic administration studies of the National Toxicology Program.
bNumber of tumors positive in NIH 3T3 transfection assays/number of tumors examined.
cUnk.: Unknown

point mutations in these H-ras genes were identified by selective oligonucleotide hybridization or by detection of a new XbaI restriction site, which was created by an AT->TA transversion at the second position of codon 61 (Table 2). Since the mutation patterns observed for each carcinogen were unique, these results are consistent with the hypothesis that these lesions were the direct result of chemical modification of H-ras gene. Because these mice only received a single dose of carcinogen shortly after birth, these observations were also consistent with ras activation as an early event in this model of hepatocarcinogenesis. Buchmann and coworkers [9] have recently confirmed the latter conclusion by demonstrating that a significant percentage of preneoplastic enzyme-altered liver lesions from diethylnitrosamine-treated B6C3F1 mice contained H-ras mutations in codon 61.

Similar conclusions were reached when activating mutations in the H-ras genes of B6C3F1 liver tumors that occurred spontaneously in two-year-old mice were compared with those of tumors induced by long-term administration of furan or furfural [37]. While all of the H-ras mutations in tumors from untreated animals were distributed among three base substitutions in codon 61, almost half of the activated H-ras genes from liver tumors induced by furan and furfural contained novel mutations outside of codon 61, which had not been previously detected in vivo. In addition, three activated K-ras genes were detected in the chemically induced tumors. These results, together with the absence of histopathological evidence for liver cytoxicity in prechronic studies at carcinogenic doses, led to the conclusion that the carcinogenic activity of furan and furfural must at least in part be due to direct mutagenic effects even though these compounds are negative in short-term bacterial mutagenicity assays.

When these studies were extended to hepatocarcinogenesis in Fischer 344 rats, a very different picture emerged. In conjunction with diethylnitrosamine-induced mouse liver tumors [41], a series of rat liver tumors induced either by continuous administration of diethylnitrosamine or by initiation with diethylnitrosamine followed by promotion with phenobarbital or 2,3,7,8-tetrachlorodibenzodioxin were screened in the NIH/3T3 assay. In sharp contrast to results with the mouse, only one of 26 rat liver tumors had transforming activity, and this tumor did not contain an activated ras gene. Comparable results have been obtained by another laboratory with a larger set of rat liver tumors induced by a similar protocol [47]. However, studies from two other laboratories have suggested that DNAs from a number of rat liver tumors were capable of inducing morphological transformation in NIH/3T3 cells. Activated K-ras genes were detected in two tumors but in most cases this transforming activity was due to unidentified non-ras genes [21,27]. Whatever the resolution of these conflicting results, it is clear that hepatocarcinogenesis in the rat,

unlike the mouse, follows a different pathway that only
rarely involves the mutational activation of ras proto-
oncogenes.

The status of ras activation in human liver cancer is
also unclear, although there have been preliminary reports
[30,51] suggesting that N-ras genes may be activated in
several tumors and cell lines derived from hepatocellular
carcinomas. In addition, a novel proto-oncogene named lca
has been cloned from a human hepatocellular carcinoma [31];
to date this gene has not been implicated in experimental
hepatocarcinogenesis. In summary, these liver tumor
studies serve as an important reminder that risk assessment
involving extrapolation between species requires careful
consideration and cannot be made solely on the basis of
ras gene activation.

Frequent activation of the K-ras proto-oncogene during
lung carcinogenesis has been detected in analogous studies
with mice and rats. Stowers et al. [40] examined a series
of lung tumors (primarily adenocarcinomas) from B6C3F1 mice
and Fischer 344 rats that were induced by long-term
exposure to tetranitromethane. Activated K-ras genes were
detected using the NIH/3T3 assay and the activating
mutations were identified by selective oligonucleotide
hybridization. All but one of these tumors contained a
GC->AT transition at the second position of the 12th codon
(Table 3). These rodent lung tumors were induced under
conditions that were comparable to threshold limit values
for human occupational exposures. Since activated K-ras
genes are fairly common in human lung adenocarcinomas, it
would be quite intriguing to look for the GC->AT mutation
in lung tumors of workers who have been exposed to this
compound. Unfortunately it was difficult to draw
mechanistic conclusions concerning the origin of the
consistent GC->AT mutations in these rodent tumors since
the mutagenic specificity of tetranitromethane was unknown
and spontaneous lung tumors from these strains were not
available for comparison.

This problem was overcome by examining lung tumors from
strain A mice in a study by You et al. [53]. Strain A mice

Table 3. Activating mutations observed in the K-ras proto-oncogene of lung tumors of B6C3F1 and strain A mice and Fischer 344 rats.[a]

Treatment	Strain	Freq.[b]	Codon 12 (GGT)				Codon 61 (CAA)			Other
			TGT	CGT	GAT	GTT	CTA	CGA	CAT	
Tetranitromethane	B6C3F1	10/10			10					
	F344	18/19			18					
Benzo(a)pyrene	A/J	14/16	8		4	1				1 K-ras
Ethyl Carbamate	A/J	10/11				1	7	2		
N-Methyl-N-nitrosourea	A/J	15/15			15					
Spontaneous	A/HeN	10/11		1	3	2		2	1	1 K-ras

[a]Data taken from References 40 and 53. Tetranitromethane-induced tumors were obtained from two-year chronic inhalation studies of the National Toxicology Program. The chemically induced strain A tumors were induced in young mice by one or a limited number of carcinogen treatments and the spontaneous tumors were obtained from untreated 24- to 27-month-old mice.
[b]Number of tumors positive in NIH 3T3 transfection assays or by selective oligonucleotide hybridization/number of tumors examined.

are highly susceptible to spontaneous and chemically induced lung tumor formation. This susceptibility made it possible to compare the pattern of K-ras mutations from tumors induced by benzo(a)pyrene, ethyl carbamate, and N-methyl-N-nitrosourea with tumors from untreated animals (Table 3). The mutation spectra in the K-ras genes of these tumors induced by different carcinogens were clearly distinct and consistent with the expected mutagenic specificity of the carcinogens. In the benzo(a)pyrene-induced tumors, GC->TA transversions and GC->AT transitions were detected. These were also the predominant base substitutions observed in mutagenesis studies in E. coli [14] and mammalian cells [28] with the diol-epoxide of benzo(a)pyrene. Likewise, mutagenesis studies [13,38] with N-methyl-N-nitrosourea, which forms O6-methylguanine adducts in DNA, predicted the GC->AT transitions in the K-ras gene of N-methyl-N-nitrosourea-induced lung tumors. This same base change was previously detected in the H-ras gene of a large number of N-methyl-N-nitrosourea-induced rat mammary carcinomas [54]. Less is known concerning the mutagenic specifity of ethyl carbamate, but vinyl carbamate is a likely intermediate in the metabolic activation of ethyl carbamate [50], so the observation that lung and liver tumors induced by these compounds contain identical patterns of mutation (Tables 2 and 3) also supports the proposition that electrophilic carcinogens interact directly with the ras genes in vivo.

3. Potential Role of Tumor Suppressor Genes

If the activation of a ras gene is an initial event in these chemically induced tumors, then what are the secondary events in the multistep carcinogenic process that cause the initiated cells to become fully malignant? Studies of transgenic mice containing oncogenes in their germ line have demonstrated that oncogenes play a causal role in tumor development. However, almost every example that has been examined in detail suggests that oncogene expression is not sufficient for full neoplastic transformation [22]. It is widely believed that one or more additional proto-oncogenes that can cooperate with the original transgene must be activated in the preneoplastic target cells of these transgenic mice for malignant tumors to develop. Unfortunately, with the exception of a study where accelerated mammary tumor formation was obtained by crossing MMTV-H-ras transgenic mice with MMTV-myc transgenic mice [42], there has been surprisingly little experimental evidence supporting the activation of cooperating oncogenes as a secondary event. Another candidate for these secondary events was suggested by a recent report [26] that examined fibrosarcomas of transgenic mice carrying the bovine papillomavirus type 1 genome. Consistent chromosome abnormalities were observed in these fibrosarcomas, including monosomy or translocations of chromosome 14 where the mouse retinoblastoma gene is located [43].

Although losses of heterozygosity have been observed frequently in many types of human neoplasia, very little is known about this type of genetic alterations in rodents. If nonrandom losses of heterozygosity could be detected in

specific types of mouse tumors, this would suggest the
presence of previously uncharacterized tumor suppressor
genes which may play a important role in experimental
carcinogenesis. We recently initiated RFLP studies with
B6C3F1 mouse tumors (outlined in Figure 1) to look for
evidence of tumor suppressor gene involvement. In this
technique DNA samples are digested with a restriction
enzyme (EcoRI in this example) that produces fragments of
unique size due to a difference in DNA sequence between the
parental mouse strains. After separating the restriction
fragments by agarose gel electrophoresis, the DNA is
transferred to a membrane and hybridized to a radioactively

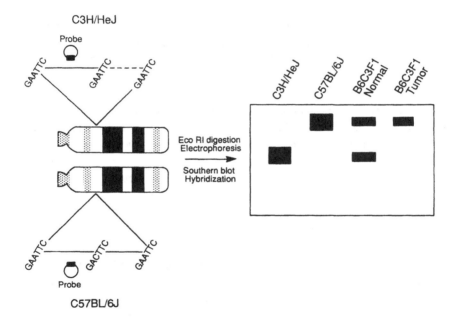

FIGURE 1. Schematic diagram outlining how an EcoRI
restriction fragment length polymorphism could be used to
detect losses of heterozygosity in tumors of B6C3F1 mice
which result from crosses between C57BL/6J and C3H/HeJ
mice (after Cavenee et al. [11]).

labelled probe. The polymorphic restriction fragments are
visualized by autoradiography and the tumor DNAs are
examined for loss of one of the progenitor alleles. Unlike
the case for humans where a particular individual is only
heterozygous for a subset of RFLP markers, the F1 offspring
of inbred mouse strains are ideal for RFLP analysis because
a panel of DNA probes can be selected that is informative
for every tumor. At present more than one hundred RFLPs
have been mapped throughout the B6C3F1 mouse genome, and
this list is growing at a rapid pace [15]. For example,
the retinoblastoma gene can be examined directly by this
technique since its cDNA detects an EcoRI RFLP on
chromosome 14 in B6C3F1 mice [43].

A large number of retroviral integration sites have
also been mapped in the mouse genome [18,19,24,44].
Integration of these retroviruses at different chromosomal
locations has created RFLPs throughout the genome which can
be detected in the F1 offspring of appropriate inbred mouse
strains. These integration sites represent a powerful
class of RFLPs since they allow multiple loci to be
simultaneously examined for losses of heterozygosity. A
series of these viral loci for B6C3F1 mice can be observed
with only five different viral DNA probes (Table 4). To
date we have examined the Emv and Mtv loci (representing
chromosomes 4, 5, 7, 8, 12, 14, and 16) in DNA from
butadiene-induced lung and liver tumors of B6C3F1 mice.
Although no differences were detected between DNA from the
liver tumors and normal livers, three of eight lung tumors
had lost the Mtv-17 locus which is located on mouse
chromosome four. A disadvantage of these viral probes is
that they only detect loss of the parental chromosome that
contains the viral integration site. Thus, losses of
heterozygosity will only be observed 50% of the time even
if one copy of a specific chromosome is lost in every
tumor. This suggests that the loss of chromosome 4 may be
a very frequent event in mouse lung tumors. We are
currently examining these tumors with additional RFLP
probes from chromosome 4 to define the minimum chromosomal
region that is lost. It is important to examine RFLP

Table 4. RFLPs due to proviral integrations in B6C3F1 mice[a]

Chromosome	Locus	Enzyme	C57BL/6J	C3H/HeJ
1	Xmv-43	EcoRI/PvuII	6.7/1.9	– / –
	Xmv-41	EcoRI/PvuII	24 /2.4	– / –
	Pmv-21	PvuII	2.2	–
2	Xmv-10	PvuII	2.2	–
	Pmv-33	PvuII	2.35	–
3	Mpmv-9	EcoRI/PvuII		
4	Pmv-30	PvuII	–	3.3
	Mtv-17	EcoRI	8.0	–
	Pmv-19	PvuII	2.4	–
	Mpmv-19	EcoRI/PvuII		
	Xmv-9	EcoRI/PvuII	3.9/4.3	– / –
	Xmv-8	EcoRI	4.4	–
	Xmv-14	EcoRI	2.55	–
	Xmv-44	EcoRI	6.0	–
	Pmv-25	EcoRI/PvuII	– / –	6.4/15
5	Mpmv-13	EcoRI/PvuII		
	Emv-1	EcoRI/PvuII	– / –	19 /4.3
	Pmv-11	EcoRI/PvuII	5.0/4.2	– / –
	Xmv-28	EcoRI/PvuII	– / –	7.1/3.7
	Pmv-12	PvuII	3.8	–
6	Xmv-24	EcoRI	–	3.7
7	Mtv-1	EcoRI	–	4.5
8	Emv-2	EcoRI/PvuII	24 /5.2	–
9	Xmv-16	PvuII	2.7	–
	Xmv-15	EcoRI	2.35	–
10	Mpmv-5	EcoRI/PvuII		
11	Pmv-2	EcoRI/PvuII	14 /9.0	– / –
	Pmv-22	EcoRI	7.1	–
	Xmv-42	EcoRI/PvuII	6.9/2.9	– / –
12	Pmv-37	EcoRI/PvuII	– / –	2.5/ 4.0
	Pmv-3	EcoRI/PvuII	4.1/8.5	– / –
	Mtv-9	EcoRI	10	–
13	Xmv-13	EcoRI/PvuII	3.0/3.5	– / –
14	Mtv-11	EcoRI	–	5.8
	Xmv-19	EcoRI/PvuII	2.1/3.6	– / –
15	Pmv-17	EcoRI	3.1	–
	Pmv-36	EcoRI/PvuII	– / –	25 /4.5
16	Mtv-6	EcoRI	–	16.7
	Pmv-35	PvuII	–	4.8
	Pmv-16	PvuII	2.55	–
17	–			
18	Xmv-29	EcoRI/PvuII	– / –	4.8/2.26
19	Xmv-18	EcoRI/PvuII	2.15/6.6	– / –

[a]Data compiled from References 18, 19, 24, and 44.

markers on the remaining mouse chromosomes in these and
additional mouse lung tumors since frequent losses of
several different chromosomes (including 3, 11, 13, and 17)
have been detected in various types of human lung cancer
[49,52].

4. Discussion

For cancer risk assessment to be truly meaningful, we
need to elucidate the molecular basis of the multiple
heritable alterations that are required for carcinogenesis.
Once these events are identified, then the mechanisms by
which chemicals affect them must be characterized. During
the past several years, strong evidence has accumulated
from the analysis of human and experimental tumors that ras
activation is one such alteration. Carcinogenesis
experiments with single doses of carcinogen [35,50,53,54]
have demonstrated that chemicals can induce ras gene
activation directly, and these mutations are likely to be
initiating events in several systems. In the initial
studies where ras mutations in tumors from long-term
carcinogen bioassays have been characterized [37,40], the
data suggest that these lesions are also induced by the
chemical treatment. Since the activated ras genes are
detected in both adenomas and carcinomas in each case,
these results also imply that the mutations occur
relatively early in the carcinogenic process. However, in
other situations such as the neoplastic transformation of
normal cells in vitro [12,20], ras activation appears to
represent a late event that results from spontaneous errors
during DNA replication. It will be interesting to extend
this type of analysis to additional chemically induced
tumors from a variety of tissues. Recent advances [1,7]
that allow ras mutations to be identified in archival
paraffin-embedded tissue sections of tumors and preneoplastic lesions using the polymerase chain reaction
technique should greatly benefit this area of research.

Other than ras proto-oncogene activation, our
understanding of chemically induced heritable alterations
during the multistep carcinogenic process is quite limited
at the molecular level. For instance, Vesselinovitch and

Michailovich [46] have suggested that at least four
critical events are required to induce hepatocellular
carcinomas by a single dose of diethylnitrosamine in
neonatal B6C3F1 mice based on a mathematical analysis of
the dose-time-response kinetics for tumor formation.
Mutations in the H-ras gene are likely to represent one of
these events in some but not all tumors, and the molecular
basis for other events during hepatocarcinogenesis remains
an important question for future research. At present the
best candidate for these additional alterations is the loss
or inactivation of genes that suppress tumorigenesis. As
we extend our RFLP studies of mouse tumors, it will be very
interesting to look for reductions to homozygosity in
chromosomal regions that contain the mouse homologs of
genes which are altered in various human tumors [34]. This
approach for tumor suppressor gene mapping in the mouse has
tremendous potential for increasing our understanding of
this important class of genes in experimental carcinogenesis. Likewise, it will be important to look for
mutations in specific genes, such as Rb and p53, that have
been identified as critical targets in human studies.

The complexity of interactions between various events
in the multistep carcinogenic process should caution
against the use of risk assessment models based on a single
genotoxic or epigenetic mechanism of action since few
chemicals, if any, will affect only one stage of
carcinogenesis when chronic exposures are involved. For
instance, the carcinogenic potency of tetranitromethane is
almost certainly a composite of its mutagenic activity and
its irritant properties. In addition, it is important to
recognize that different types of genetic damage may be
important for oncogenes and tumor suppressor genes.
Examples of proto-oncogene activation have been reported to
involve point mutations, translocations, and gene
amplification [6], while tumor suppressor genes in contrast
may be mutated or inactivated by deletions, chromosome
nondisjunction, somatic recombination, gene conversion, or
point mutations [11]. Thus chemicals that lack detectable
activity in gene mutation assays may still exhibit

significant genetic activity by inducing chromosomal
mutations such as nondisjunction or deletion.

Acknowledgements

The author would like to thank Carl Barrett and Jill
Stowers-Hoffman for helpful comments and Sandy Sandberg
for preparation of the manuscript.

References

1. Almoguera, C., D. Shibata, K. Forrester, J. Martin, N.
    Arnheim, and M. Perucho. Most human carcinomas
    of the exocrine pancreas contain mutant c-K-ras
    genes. Cell 53: 549-554, 1988.

2. Baker, S.J., E.R. Fearon, J.M. Nigro, S.R. Hamilton,
    A.C. Preisinger, J.M. Jessup, P. Vantuinen, D.H.
    Ledbetter, D.F. Barker, Y. Nakamura, R. White,
    and B. Vogelstein. Chromosome 17 deletions and
    p53 gene mutations in colorectal carcinomas.
    Science 244: 217-221, 1989.

3. Balmain, A. and K. Brown. Oncogene activation in
    chemical carcinogenesis. Adv. Cancer Res. 51:
    147-182, 1988.

4. Barbacid, M. ras oncogenes. Annu. Rev. Biochem.
    56: 779-827, 1987.

5. Barrett, J.C. and R.W. Wiseman. Cellular and molecu-
    lar mechanisms of multistep carcinogenesis:
    relevance to carcinogen risk assessment.
    Environ. Health Perspect. 76: 65-70, 1987.

6. Bishop, J.M. The molecular genetics of cancer.
    Science 235: 303-311, 1987.

7. Bos, J.L. ras oncogenes in human cancer: a review.
    Cancer Res. 49: 4682-4689, 1989.

8. Bos, J.L., E.R. Fearon, S.R. Hamilton, M.V. de Vries,
    J.H. van Boom, A.J. van der Eb, and B.
    Vogelstein. Prevalence of ras gene mutations in
    human colorectal cancers. Nature (London) 327:
    293-297, 1987.

9. Buchmann, A., J. Mahr, R. Bauer-Hofmann, and M.
    Schwarz. Mutations at condon 61 of the
    Ha-ras proto-oncogene in precancerous liver
    lesions of the B6C3F1 mouse. Molec. Carcinogen.
    2: 121-125, 1989.

10. Cavenee, W.K., T.P. Dryja, R.A. Phillips, W.F.
    Benedict, R. Godbout, B.L. Gallie, A.L.
    Murphree, L.C. Strong, and R.L. White.
    Expression of recessive alleles by chromosomal
    mechanisms in retinoblastoma. Nature (London)
    305: 779-784, 1983.

11. Cavenee, W.K., A. Koufos, and M.F. Hansen. Recessive mutant genes predisposing to human cancer. *Mutat. Res.* 168: 3-14, 1986.

12. Doniger, J., V. Notario, and J.A. DiPaolo. Carcinogens with diverse mutagenic activities initiate neoplastic guinea pig cells that acquire the same N-ras point mutation. *J. Biol. Chem.* 262: 3813-3819, 1987.

13. DuBridge, R.B., P. Tang, H.C. Hsia, P-M. Leong, J.H. Miller, and M.P. Calos. Analysis of mutation in human cells by using an Epstein-Barr virus shuttle system. *Mol. Cell. Biol.* 7: 379-387, 1987.

14. Eisenstadt, E., A.J. Warren, J. Porter, D. Atkins, and J.H. Miller. Carcinogenic epoxides of benzo[a]pyrene and cyclopenta[cd]pyrene induce base substitutions via specific transversions. *Proc. Natl. Acad. Sci. USA* 79: 1945-1949, 1982.

15. Elliott, R. DNA restriction fragment variants. *Mouse News Lett.* 83: 126-148, 1989.

16. Finlay, C.A., P.W. Hinds, and A.J. Levine. The p53 proto-oncogene can act as a suppressor transformation. *Cell* 57: 1083-1093, 1989.

17. Forrester, K., C. Almoguera, K. Han, W.E. Grizzle, and M. Perucho. Detection of high incidence of K-ras oncogenes during human colon tumorigenesis. *Nature* (London) 327: 298-303, 1987.

18. Frankel, W.N., J.P. Stoye, B.A. Taylor, and J.M. Coffin. Genetic analysis of endogenous xenotropic murine leukemia viruses: association with two common mouse mutations and the viral restriction locus Fv-1. *J. Virol.* 63: 1763-1774, 1989.

19. Frankel, W.N., J.P. Stoye, B.A. Taylor, and J.M. Coffin. Genetic identification of endogeneous polytropic proviruses by using recombinant inbred mice. *J. Virol.* 63: 3810-3821, 1989.

20. Gilmer, T.M., L.A. Annab, and J.C. Barrett. Characterization of activated proto-oncogenes in chemically transformed Syrian hamster embryo cells. *Mol. Carcinogen.* 1: 180-188, 1988.

21. Goyette, M., M. Dolan, W. Kaufmann, D. Kaufman, P.R. Shank, and N. Fausto. Transforming activity of DNA from rat liver tumors induced by the carcinogen methyl(acetoxymethyl)nitrosamine. *Mol. Carcinogen.* 1: 26-32, 1988.

22. Hanahan, D. Dissecting multistep tumorigenesis in transgenic mice. *Annu. Rev. Genet.* 22: 479-519, 1988.

23. Huff, J. Long term in vivo results for chemical carcinogens. *Proc. Am. Assoc. Cancer Res.* 30: 447, 1989.

66

24. Jenkins, N.A., N.G. Copeland, B.A. Taylor, H.G. Bedigian, and B.K. Lee. Ecotropic murine leuke- mia virus DNA content of normal and lymphomatus tissues of BXH-2 recombinant inbred mice. J. Virol. 42, 379-388, 1982.

25. Klein, G. The approaching era of the tumor suppressor genes. Science 238: 1539-1545, 1987.

26. Lindgren, V., M. Sippola-Thiele, J. Skowronski, E. Wetzel, P.M. Howley, and D. Hanahan. Specific chromosomal abnormalities characterize fibrosar- comas of bovine papillomavirus type 1 transgenic mice. Proc. Natl. Acad. Sci. USA 86: 5025-5029, 1989.

27. McMahon, G., L. Hanson, J-J. Lee, and G.N. Wogan. Identification of an activated c-Ki-ras oncogene in rat liver tumors induced by aflatoxin B_1. Proc. Natl Acad. Sci. USA 83: 9418-9422, 1986.

28. Mazur, M. and B.W. Glickman. Sequence specificity of mutations induced by benzo[a]pyrene-7,8-diol-9, 10-epoxide at endogenous aprt gene in CHO cells. Somat. Cell. Mol. Genet. 14: 393-400, 1988.

29. Newcomb, E.W., J.J. Steinberg, and A. Pellicer. ras Oncogenes and phenotypic staging in N-methyl- nitrosourea- and γ-irradiation-induced thymic lymphomas in C57BL/6J mice. Cancer Res. 48: 5514-5521, 1988.

30. Notario, V., S. Sukumar, E. Santos, and M. Barbacid. A common mechanism for the malignant activation of ras oncogenes in human neoplasia and in chemically induced animal tumors, in: Cancer Cells, Oncogenes and Viral Genes (G.F. Vande Woude, A.J. Levine, W.C. Topp, and D. Watson, eds.), pp. 425-432, Cold Spring Harbor Laboratory, Cold Spring Harbor, NY, 1984.

31. Ochiya, T., A. Fujiyama, S. Fukushige, I. Hatada, and K. Matsubara. Molecular cloning of an oncogene from a human hepatocellular carcinoma. Proc. Natl. Acad. USA 83: 4993-4997, 1986.

32. Parada, L.F. and R.A. Weinberg. Presence of a Kirsten murine sarcoma virus ras oncogene in cells transformed by 3-methylcholanthrene. Mol. Cell. Biol. 3: 2298-2301, 1983.

33. Perantoni, A.O., J.M. Rice, C.D. Reed, M. Watatani, and M.L. Wenk. Activated neu oncogene sequences in primary tumors of the peripheral nervous system induced in rats by transplacental expo- sure to ethylnitrosourea. Proc. Natl. Acad. Sci. USA 84: 6317-6321, 1987.

34. Ponder, B. Gene losses in human tumors. Nature (London) 335: 400-402, 1988.

35. Quintanilla, M., K. Brown, M. Ramsden, and A. Balmain. Carcinogen specific mutation and

amplification of Ha-ras during mouse skin carcinogenesis. Nature (London) 322: 78-80, 1986.

36. Reynolds, S.H., S.J. Stowers, R.P. Maronpot, M.W.
    Anderson, and S.A. Aaronson. Detection and
    identification of activated oncogenes in spon-
    taneously occurring benign and malignant hepato-
    cellular tumors of the B6C3F1 mouse. Proc.
    Natl. Acad. Sci. USA 83: 33-37, 1986.

37. Reynolds, S.H., S.J. Stowers, R.M. Patterson, R.P.
    Maronpot, S.A. Aaronson, and M.W. Anderson.
    Activated oncogenes in B6C3F1 mouse liver
    tumors: implications for risk assessment.
    Science 237: 1309-1316, 1987.

38. Richardson, K.K., F.C. Richardson, R.M. Crosby, J.A.
    Swenberg, and T.R. Skopek. DNA base changes and
    alkylation following in vivo exposure of
    Escherichia coli to N-methyl-N-nitrosourea of
    N-ethyl-N-nitrosourea. Proc. Natl. Acad. Sci.
    USA 84: 344-348, 1987.

39. Rodenhuis, S., R.J.C. Slebos, A.J.M. Boot, S.G.
    Evers, W.J. Mooi, S.Sc. Wagenaar, P.Ch. van
    Bodegom, and J.L. Bos. Incidence and possible
    clinical significance of K-ras oncogene
    activation in adenocarcinoma of the human lung.
    Cancer Res. 48: 5738-5741, 1988.

40. Stowers, S.J., P.L. Glover, S.H. Reynolds, L.R.
    Boone, R.P. Maronpot, and M.W. Anderson.
    Activation of the K-ras proto-oncogene in lung
    tumors from rats and mice chronically exposed to
    tetranitromethane. Cancer Res. 47, 3212-3219,
    1987.

41. Stowers, S.J., R.W. Wiseman, J.M. Ward, E.C. Miller,
    J.A. Miller, M.W. Anderson, and A. Eva.
    Detection of activated proto-oncogenes in N-
    nitrosodiethylamine-induced liver tumors: a com-
    parison between B6C3F1 mice and Fischer 344
    rats. Carcinogenesis 9: 271-276, 1988.

42. Sinn, E., W. Muller, P. Pattengale, I. Tepler, R.
    Wallace, and P. Leder. Coexpression of
    MMTV/v-ras and MMTV/c-myc genes in transgenic
    mice: synergistic action of oncogenes in vivo.
    Cell 49: 465-475, 1987.

43. Stone, J.C., J.L. Crosby, C.A. Kozak, A.R.
    Schievella, R. Bernards, and J.H. Nadeau. The
    murine retinoblastoma homolog maps to chromosome
    14 near Es-10. Genomics 5: 70-75, 1989.

44. Traina, V.L., B.A. Taylor, and J.C. Cohen. Genetic
    mapping of endogenous mouse mammary tumor viru-
    ses: locus characterization, segregation, and
    chromosomal distribution. J. Virol. 40:
    735-744, 1981.

45. Vogelstein, B., E.R. Fearon, S.R. Hamilton, S.E.
    Kern, A.C. Preisinger, M. Leppert, Y. Nakamura,

68

R. White, A.M.M. Smits, and J.L. Bos. Genetic alterations during colorectal-tumor development. N. Engl. J. Med. 319: 525-532, 1988.

46. Vesselinovitch, S.D. and N. Mchailovich. Kinetics of diethylnitrosamine hepatocarcinogenesis in the infant mouse. Cancer Res. 43: 4253-4259, 1983.

47. Watatani, M., A.O. Perantoni, C.D. Reed, T. Enomoto, M.L. Wenk, and J.M. Rice. Infrequent activation of K-ras, H-ras, and other oncogenes in hepato-cellular neoplasms initiated by methyl(acetoxy-methyl)-nitrosamine, a methylating agent, and promoted by phenobarbital in F344 rats. Cancer Res. 49: 1103-1109, 1989.

48. Weinberg, R.A. Oncogenes, antioncogenes, and the molecular basis of multistep carcinogenesis. Cancer Res. 49: 3713-3721, 1989.

49. Weston, A., J.C. Willey, R. Modali, H. Sugimura, E.M. McDowell, J. Resau, B. Light, A. Haugen, D.L. Mann, B.F. Trump, and C.C. Harris. Differen-tial DNA sequence deletions from chromosomes 3, 11, 13, and 17 in squamous-cell carcinoma, large-cell carcinoma, and adenocarcinoma of the human lung. Proc. Natl. Acad. Sci. USA 86: 5099-5103, 1989.

50. Wiseman, R.W., S.J. Stowers, E.C. Miller, M.W. Anderson, and J.A. Miller. Activating mutations of the c-Ha-ras proto-oncogene in chemically induced hepatomas of the male B6C3F1 mouse. Proc. Natl. Acad. Sci. USA 83: 5825-5829, 1986.

51. Yager, J.D. and J. Zurlo. Oncogene activation and expression during carcinogenesis in liver and pancreas, in: The Pathobiology of Neoplasia (A.E. Sirica, Ed.), Plenum Publishing Corp., pp. 399-415, 1989.

52. Yokota, J., M. Wada, Y. Shimosato, M. Terada, and T. Sugimura. Loss of heterozygosity on chromosomes 3, 13, and 17 in small-cell carcinoma and on chromosome 3 in adenocarcinoma of the lung. Proc. Natl. Acad. Sci. USA 84: 9252-9256, 1987.

53. You, M., U. Candrian, R.P. Maronpot, G.D. Stoner, and M.W. Anderson. Activation of the K-ras proto-oncogene in spontaneously occurring and chemi-cally induced lung tumors of the strain A mouse. Proc. Natl. Acad. Sci. USA 86: 3070-3074, 1989.

54. Zarbl, H., S. Sukumar, A.V. Arthur, D. Martin-Zanca and M. Barbacid. Direct mutagenesis of Ha-ras-1 oncogenes by N-nitroso-N-methylurea during ini-tiation of mammary carcinogenesis in rats. Nature (London) 315: 382-385, 1985.

Multistage Hepatocarcinogenesis in the Rat as a Basis for Models of Risk Assessment of Carcinogenesis

Henry C. Pitot, Mark J. Neveu, James R. Hully,
Tahir A. Rizvi, and Harold Campbell

Although biological processes are, in general, far more complex than chemical and physical phenomena, one goal of biology, toxicology, and pathology is the mathematical-statistical formulation of models that describe the mechanisms of normal and disease processes. However, as pointed out by several scientists (Whittemore, 1978; Moolgavkar, 1986; Alavanja et al., 1987), up to this time no mathematical-statistical model of a biological process has faithfully described all of the particulars of the biological phenomenon under consideration. This is especially true in the field of carcinogenesis (cf. Chu, 1987) where many models have been proposed in the past. With any model, however, its formulation is dependent on the level of knowledge of the biological mechanisms controlling the process, i.e., carcinogenesis. Recent advances in our understanding of the mechanisms of carcinogenesis as a multistage process have allowed, at least potentially, for a closer congruence between models representative of carcinogenesis and their pathogenesis.

A.   Multistage Models

While the multistage nature of cancer development has
been known for almost five decades (cf. Scribner and Süss,
1978), a number of mathematical-statistical models of carcinogenesis proposed during this time period did not take
into account the multistage nature of carcinogenesis.
These include the classical one-hit (linear) model, the
multi-hit (k-hit) model, the logistic model, the probit
model, and the Weibull model (cf. Gaylor and Shapiro, 1979;
Hanes and Wedel, 1985). The reader is directed to the references for the specific formulation of these models. In
general, these models reflect the concept of a non-
threshold linear extrapolation at low doses of carcinogen.
It is this latter characteristic that has pervaded the concepts on which regulatory decisions are made concerning
carcinogenic agents. However, none of these models can be
shown to reflect accurately the biological phenomenon of
carcinogenesis, although some may be more closely aligned
with the biological characteristics of carcinogenesis than
others. Clearly, attempts to simply match data results
with one or another of these models, making various assumptions to account for failures of the model, are not an
appropriate method of risk estimation.

In order to relate mathematical-statistical models of
carcinogenesis more closely to the biological phenomenon
itself, it is appropriate that models of carcinogenesis
take into account the multistage nature of the development
of neoplasia; this concept is rapidly becoming generally
accepted for all histogenetic types of neoplasms (cf.
Pitot, 1989). Among the first attempts to develop a bio-
mathematical representation of multistage carcinogenesis
were the proposals by Muller (1951) and Nordling (1953);
however, certain predictions of these models were in clear
disagreement with the results of some carcinogenesis experiments (cf. Chu, 1987). In 1954, Armitage and Doll proposed their multistage model, which simply indicates that a

normal cell must go through a series of stages or changes to become a transformed or malignant cell. This model reflects most closely the biological concepts of Foulds (1954), who termed all changes beyond the initiation (see below) of a cell as reflecting continuous but multiple changes to malignancy. The Armitage and Doll model makes several predictions that do coincide with epidemiologic and experimental data. These include the pattern of increase in cancer incidence with age seen in the human. In addition, the model can make some predictions with respect to dose response of chemical carcinogens on the assumption that the transition rate from one stage to another is a linear function of the dose of the carcinogen. In this instance the cancer incidence would be a linear function of the dose of the carcinogen when only one transition rate is affected. If two or more are affected, then the incidence becomes a quadratic or higher function of the carcinogenic dose.

However, the Armitage and Doll model does not take into account certain biological features now known concerning specific stages in multistage carcinogenesis. Specifically, since cell kinetics are not explicitly considered, the stage of promotion (see below) cannot be adequately modeled.

A model that more directly reflects the multistage nature of neoplastic development has been proposed by Moolgavkar and his associates (cf. Moolgavkar, 1986). A diagram of this model is seen in Figure 1. This model can account for much of the epidemiology of human cancer. Furthermore, it postulates at least three stages in the development of neoplasia, as discussed below. As indicated above, our understanding of multistage carcinogenesis has rapidly advanced within the past decade so that a detailed analysis of the biology of at least one specific system, multistage hepatocarcinogenesis in the rat, can be made in relation to the multistage model of carcinogenesis seen in Figure 1.

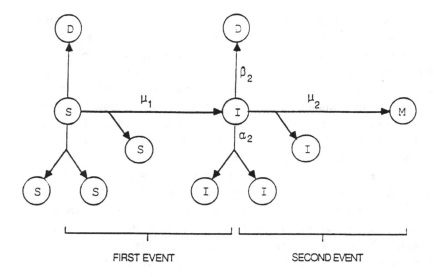

FIGURE 1. Moolgavkar representation of multistage carcinogenesis. S = normal stem cells, I = intermediate (one-
hit) cell, D = differentiated or deceased cell, M = malignant cell; μ_1 = rate of first event occurrence, μ_2 = rate
of second event occurrence, α_2 = rate of division of intermediate cells (cells in the stage of promotion), β_2 = rate
of differentiation and death of cells in the stage of
promotion. Adapted from Moolgavkar (1986).

B. Multistage Hepatocarcinogenesis in the Rat

As indicated above, the multistage nature of the natural history of the development of the neoplastic process
has been known for almost five decades. The history of the
development of our knowledge in this area is beyond the
scope of this discussion, but the reader is referred to
several reviews (Scribner and Süss, 1978; Farber and
Cameron, 1980; Slaga, 1983). These reviews consider the
natural history of neoplastic development in two stages,
termed initiation and promotion, the latter always following the former. More recently a third stage, termed progression, in part because of its similarity to the use of
the term by Foulds (1954) in the context indicated above,

has been identified. The characterization, quantitation, and usefulness of an understanding of these three stages in hepatocarcinogenesis are to be discussed. The restriction of the discussion to the development of a single histogenetic type of neoplasm herein is more for convenience than for exclusion, since it is quite likely that other histogenetic models of multistage carcinogenesis could be exploited in a similar manner.

    1.   Characteristics of the stages of hepatocarcinogenesis in the rat.

In Table 1 are listed the predominant characteristics of the stages of initiation, promotion, and progression in hepatocarcinogenesis.

In the liver as in the skin, the process of initiation, the first stage in the natural history of neoplastic development, is permanent and irreversible, as demonstrated by the extended time that may occur between initiation and subsequent promotion with virtually no loss in the appearance of lesions (Pitot, 1979). However, the effectiveness of initiation appears to depend on its relationship in time to cellular replicative DNA synthesis and cell division (Ying et al., 1982; Ishikawa et al., 1980). Furthermore, the process of DNA synthesis itself is critical for the "fixation" and thus the irreversibility of initiation (Warwick, 1971). In view of the fact that "spontaneous" neoplasms occur ubiquitously both in the human and experimental animals, one must conclude that "spontaneous" initiation must have occurred as the first stage in the development of such neoplasms. Hepatocytes with the characteristics of "spontaneous" or "fortuitous" initiated cells have been identified in the livers of rats (Emmelot and Scherer, 1980; Popp et al., 1985) and the mouse (Lee et al., 1989). As the age of the animal increases, the total number of altered hepatocytes developing from such spontaneous initiation also increases (Schulte-Hermann et al.,

Table 1. Biological Characteristics of the Stages of Initiation, Promotion, and Progression in Hepatocarcinogenesis[1]

Initiation	Promotion	Progression
Irreversible, with constant "stem cell" potential	Reversible increase in replication of progeny of the initiated cell population	Irreversible. Measurable and/or morphologically discernible alteration in cell genome's structure
Efficacy sensitive to xenobiotic and other chemical factors	Reversible alterations in gene expression	Growth of altered cells sensitive to environmental factors during early phase
Spontaneous (fortuitous) occurrence of initiated cells can be demonstrated	Promoted cell population existence dependent on continued administration of the promoting agent	Benign and/or malignant neoplasms characteristically seen
Requires cell division for "fixation"	Efficacy sensitive to dietary and hormonal factors	"Progressor" agents act to advance promoted cells into this stage but may not be initiating agents
Dose response does not exhibit a readily measurable threshold	Dose response exhibits measurable threshold and maximal effect dependent on dose of initiating agent	Spontaneous (fortuitous) progression can be demonstrated
Relative effect of initiators depends on quantitation of focal lesions following defined period of promotion	Relative effectiveness of promoters depends on their ability with constant exposure to cause an expansion of the progeny of the initiated cell population	

[1]Adapted from Pitot et al. (1989).

1983; Ogawa et al., 1981), although the number of initiated
cell clones tends to level off as the animal grows older
(Pitot et al., 1985). The effectiveness of initiation can
also be altered as a result of the environmental modification of the metabolism of initiating agents to their
ultimate reactive forms (cf. Wattenberg, 1978).

The stage of promotion is clearly distinguished from
the stages of initiation and progression by its reversibility or instability. Such reversibility can be readily
demonstrated in multistage hepatocarcinogenesis (Bursch et
al., 1984; Hendrich et al., 1986). Furthermore, focal
lesions (altered hepatic foci) that disappear when application of the promoting agent is discontinued will reappear
on readministration of the promoting agent (Hendrich et
al., 1986). The exact mechanism of the loss of these cell
populations during the period of promotion after removal of
the promoting agent is not entirely clear, but there is
significant evidence that individual cell death plays a
major role (Bursch et al., 1984). In concert with these
findings, studies by Hanigan and Pitot (1985) demonstrated
that hepatocytes already in the stage of promotion are
dependent on the presence of the promoting agent *in vivo*
for their continued existence. This phenomenon is completely analogous to the "dependent" neoplasms of endocrine
tissues described more than two decades ago by Furth
(1968).

From studies primarily in multistage carcinogenesis in
mouse epidermis, Boutwell (1974) proposed that the principal molecular action of promoting agents is their alteration of genetic expression, an effect reversed on withdrawal of the promoting agent. Many promoting agents,
including those relatively specific for tumor promotion in
hepatocytes, exert their effects on gene expression by
several mechanisms such as through the mediation of receptors (cf. Pitot et al., 1988). The regulation of the
expression of several genes is altered by the chronic
administration of promoting agents for the liver, including

enzymes of xenobiotic metabolism (cf. Williams, 1984), γ-glutamyltranspeptidase (Kitagawa et al., 1980; Sirica et al., 1984), and the placental form of glutathione S-transferase (Dock et al., 1984). Unlike initiation, the stage of promotion can be continually modulated by a variety of environmental alterations, which include the composition and amount of the diet (Hendrich et al., 1988) as well as specific chemical agents including vitamin E (Ura et al., 1987) and aminophenols (Kurata et al., 1985). Dose response characteristics of initiating and promoting agents will be considered below.

With a clear demonstration of the stages of initiation and promotion as critical components in the development of hepatic neoplasia and the reversible nature of the stage of promotion, it is clear that the appearance of irreversible, malignant neoplasms represents a distinct stage in this process (cf. Schulte-Hermann, 1985; Pitot et al., 1988). From the characteristics of the irreversible stage of progression listed in Table 1, readily demonstrable changes in the structure of the genome of the neoplastic cell are the hallmark of this stage. Such changes can be directly related to the increased growth rate, invasiveness, metastatic capability, and biochemical changes that occur in the neoplastic cell during this stage. These latter characteristics of the stage of progression were emphasized by Foulds (1965), but a greater understanding of the relationship of such changes to molecular alterations in the cell genome can now be made. In hepatocarcinogenesis, aneuploidy of hepatocellular carcinomas has been previously described (Nowell and Morris, 1969), as well as aneuploidy in focal lesions induced under conditions in which such cells would be in the stage of progression (Sargent et al., 1989).

Within the last decade, following the suggestion by Potter (1981) and Moolgavkar and Knudson (1981), formats for multistage carcinogenesis have adapted the initiation-promotion-initiation protocol, with the last initiation actually being the entrance into the stage of progression. Such a format leads to a marked increase in the number of carcinomas, whereas the simple initiation-promotion format results predominantly in benign neoplasms. This phenomenon is true both for multistage carcinogenesis in the mouse epidermis (Hennings et al., 1985) and in multistage hepato-carcinogenesis in the rat (Scherer, 1984; Pitot et al., unpublished observations). In multistage hepatocarcino-genesis in the rat, such a format results in numerous altered hepatic foci, a number of which contained foci of biochemically and morphologically distinguishable cells, so-called "foci-in-foci" (Scherer, 1984). Such morphological alterations have been suggested to represent the earliest discernible cells in the stage of progression in liver (Scherer, 1987; Hirota and Yokoyama, 1985). The occurrence of foci-in-foci in a standard initiation-promotion protocol of rat hepatocarcinogenesis is relatively infrequent (Pitot et al., 1978). In fact, it is likely that such heterogeneous lesions reflect spontaneous (fortuitous) progression in such lesions. Furthermore, the initiation-promotion-initiation protocol can be utilized to identify putative "progressor" agents. Such agents by themselves would be capable only of converting cells in the stage of promotion to the stage of progression. Obviously, complete carcinogens exhibit progressor agent activity, since they are capable of inducing the entire process of carcinogenesis. To date there have been no distinct examples of agents exhibiting only progressor activity, but in view of the characteristics of the stage of progression, especially those dealing with karyotypic alterations, one might propose that some clastogenic agents might exhibit progressor activity without demonstrable initiating and/or promoting action.

2.   Qualitative and quantitative analysis of the
stages of hepatocarcinogenesis in the rat.

Since it has been possible to characterize the three
stages of hepatocarcinogenesis in the rat in relation to
certain biological parameters, a logical extension of these
studies is the quantitation of alterations seen in these
three stages. In general, this has been done or at least
can potentially be carried out for each of the three
stages.

Quantitation of the stage of initiation is usually
performed by using phenotypic markers to enumerate the number of small focal lesions that are induced by an initiating agent followed by the administration of a promoting
agent, resulting in clonal expression in initiated cells to
form foci that are readily measured and quantified
(Goldsworthy et al., 1984). Quantitation of the stage of
promotion involves a determination of changes in the total
number of cells present in the stage of promotion at any
one timepoint. This is a direct reflection of the total
volume occupied by all lesions in the organ. Finally, the
stage of progression could theoretically be quantitated by
the determination of the number of foci-in-foci, which
herald the beginning of the stage of progression (Scherer,
1987). Unfortunately, at the present time there is no
known method to enumerate quantitatively such foci-in-foci
within a solid organ. On the other hand, it is possible to
quantitate the number of altered hepatic foci that remain
after removal of the promoting agent, since the cells of
such focal lesions appear to be independent of the presence
of the promoting agent during the stage of progression
(Hanigan and Pitot, 1985; Hendrich et al., 1986).

a.   Background parameters.

In any method of the quantitation of carcinogenesis *in
vivo*, it is critical to correct all calculations for back-

ground or spontaneous lesions. This is especially true in multistage hepatocarcinogenesis, in which background or fortuitous initiation, promotion, and progression occur by mechanisms that, to a great extent, cannot readily be controlled by experimental conditions. This phenomenon is reflected in the incidence of spontaneous neoplasia not only in experimental animals (cf. Ward, 1983), but in the human as well (Devesa et al., 1987).

In the liver of the rat, the spontaneous occurrence of altered hepatic foci (AHF), which are the clonal progeny of initiated cells (cf. Pitot et al., 1988), has been described by a number of investigators (Pollard et al., 1982; Schulte-Hermann et al., 1983; Popp et al., 1985). Furthermore, Moore et al. (1987), using a specific immuno-histochemical marker for AHF, have identified single hepatocytes exhibiting this marker, whereas normal hepatocytes do not express the marker. It is not clear, however, how many or which of these single altered hepatocytes have the potential to develop clonally under the influence of a promoting agent, as is characteristic of truly initiated hepatocytes.

The number of spontaneous altered hepatic foci in the presence of a promoting agent appears to increase until the animal is about 3 months of age, and then very little increase is noted through 11 months of age (Pitot et al., 1985). In the absence of exogenous promotion, the number of spontaneous focal lesions per liver increases dramatically between 85 and 111 weeks of age (Popp et al., 1985). Such alterations would be expected in response to uncontrolled initiating agents such as background radiation, dietary contaminants, or endogenous carcinogenic metabolites.

Since a number of endogenous factors are promoting agents, including steroid sex hormones (Taper, 1978; Wanless and Medline, 1982), some pituitary polypeptide hormones, especially prolactin (Buckley et al., 1985; Dao and Chan, 1983), and metabolites such as bile acids (Tsuda et al., 1984), hepatocytes initiated either spontaneously or by the action of exogenous carcinogens will proliferate under such influences. In fact, spontaneous foci increase in size as the animal ages, as would be expected by the action of endogenous promoting agents (Popp et al., 1985; Schulte-Hermann et al., 1983). Furthermore, crude diets, in contrast to purified diets, do exhibit promoting activity as well as synergism with exogenously added promoters (Hendrich et al., 1988). Therefore, experiments on multistage carcinogenesis, whether in the liver or other tissues, must consider promotion by endogenous and exogenous "background" promoting agents in the determination of the effectiveness and potency of an experimental or exogenous agent added to the system to test its effect.

As with initiation, spontaneous or fortuitous progression must also occur in light of the appearance of spontaneous malignant neoplasms (see above). In multistage hepatocarcinogenesis, the appearance of "foci-in-foci" has been described under conditions in which no specific agent with progressor activity has been specifically added to the system (Pitot et al., 1978). Furthermore, the continued administration of promoting agents, even in the absence of exogenous initiation, can give rise to malignant neoplasms both in the liver (Kociba et al., 1978; Rossi et al., 1977) and in epidermal carcinogenesis (Iversen, 1985). Therefore, there is ample evidence that spontaneous or fortuitous progression also exists, but because of the difficulties in the exact quantitation of this stage of neoplastic development in hepatocarcinogenesis, controls for spontaneous progression must remain rather crude.

b. Determination of the relative potencies of agents as initiators, promoters, and/or progressors.

A major variable in risk assessment that has received little attention as yet is the determination of the relative potency of carcinogenic agents, one to another. There is ample evidence that such a variation exists (cf. Ames et al., 1982), but the exactness with which such potency can be determined is quite variable. One of the potential reasons for such variation is the multistage nature of carcinogenesis. As an alternative to utilizing simply the number and time until appearance of malignant neoplasm as the critical parameters in carcinogenic potency, one may choose to determine the potency of a compound as an initiating agent, promoting agent, and/or progressor agent. Because of the relative ease of the quantitation of lesions in the stages of initiation and promotion in multistage hepatocarcinogenesis in the rat, such a system lends itself to the determination of relative potencies of agents as initiators and/or promoters. By use of the number of altered hepatic foci (AHF) as well as the volume percentage of the liver occupied by these lesions, as determined from quantitative stereologic calculations, parameters for the estimation of the relative potency of chemicals as initiating or promoting agents can be determined (Pitot et al., 1987). The parameters that have been used for estimating such relative potencies are as follows:

INITIATION INDEX = [TOTAL NUMBER OF AHF - NUMBER OF SPONTANEOUS AHF] x $LIVER^{-1}$ x $[mmol/kg BODY WEIGHT]^{-1}$

PROMOTION INDEX = [VOLUME FRACTION (%) OF LIVER OCCUPIED BY AHF IN TEST ANIMALS] ÷ [VOLUME FRACTION (%) OF LIVER OCCUPIED BY AHF IN ANIMALS INITIATED BUT NOT RECEIVING TEST AGENT] x $mmol^{-1}$ x $WEEKS^{-1}$

The initiation index must be determined from values on the linear portion of the dose-response curve, if the dose-response curve levels off because of toxicity or other undetermined reasons (cf. Swenberg et al., 1987). Thus, in theory at least, the initiation index should not vary significantly over the linear dose range. In fact, this has been reported for at least two carcinogens (Pitot et al., 1987).

The promotion index also must be calculated from the linear portion of the dose-response curve if relative potencies of agents as promoting agents are to be compared. Earlier studies with the hepatic promoting agent, phenobarbital, demonstrated a sigmoid-like curve with a relatively linear segment after a threshold and a leveling off or even downward response at high doses (Goldsworthy et al., 1984). The threshold of promoting agents has been described in a number of publications both in multistage hepatocarcinogenesis (Driver and McLean, 1986; Goldsworthy et al., 1984; Pitot et al., 1987) and multistage epidermal carcinogenesis (Verma and Boutwell, 1980). The plateau effect at higher doses is probably the result of the fact that there are a finite number of initiated cells, and when all are promoted to quantifiable AHF, a plateau effect would be expected. This point is also very significant in determination of the initiation index, since accurate determination of this parameter would require that all initiated cells be promoted to AHF that could be detected and enumerated by the techniques utilized. As with the initiation index, one might predict that as long as the promotion index is determined from values on the linear portion of the dose-response curve of a promoting agent, such values should be essentially identical. Recently, Rizvi et al. (1988) have demonstrated this to be true, noting, however, that at doses near or below the threshold value the promotion index assumes abnormally high values. Whether this represents an extremely high efficiency of promotion at very low concentrations or an artifact of the dose-response curve has not been determined as yet.

Some representative values for initiation and promotion indices for a variety of chemicals can be found in the reference by Pitot et al. (1987). From these data it can be seen that there is a considerable range of potencies both for initiation and promotion. It should also be noted that it is possible to determine both initiating and promoting potencies of a complete carcinogen by a suitable

format of the experiment, meeting the requirements as indicated above. For a different perspective on initiating and
promoting indices, see Moolgavkar et al. (this volume).

As yet, determination of the relative effectiveness of
agents as progressors has not been successful or even reasonably attempted. The principal reason for this is the
lack of accurate quantitation of the stage of progression
by present techniques. The crude, gross determination of
malignant neoplasms in a solid organ, such as the liver, is
quite unsatisfactory and inaccurate, and as yet the quantitative determination of the numbers of foci-in-foci, on the
assumption that these represent the first morphologic
development of the stage of progression, has not been successful. As indicated above, however, it may be possible
to estimate the effectiveness of progressor agents by quan-
titating the "promoter-independent" altered hepatic foci
remaining after withdrawal of the promoting agent. In the
final analysis, however, since it is the malignant neoplasm
that is of most concern clinically, the determination of
the potency of carcinogens as progressor agents may be most
significant to risk estimation. On the other hand, chemicals exhibiting only action as promoting agents should be
identified and their potency determined, since such agents
do have the capacity to promote spontaneously initiated
cells, whose clones are at a significantly higher risk for
entrance into the stage of progression.

C. **Multistage Hepatocarcinogenesis in the Rat as a Basis
    for Models of Cancer Development**

A major difficulty in relating mathematical-
statistical models of carcinogenesis to the biological
phenomenon itself has been the inability to determine many
of the parameters necessary for a number of the variables
of models that have been proposed. Therefore, the more
accurately one can determine the biological parameters of
the stages of carcinogenesis, the better the fit between

the biological parameters that can be measured and the mathematical formulations made. At the present time, multistage hepatocarcinogenesis in the rat appears to be a biological system that can potentially be used for the testing of one or more mathematical-statistical models. In this paper we have chosen to utilize the Moolgavkar-Knudson model of multistage carcinogenesis to demonstrate how it might be used to make predictions that could be developed and verified by the multistage hepatocarcinogenesis system in the rat. This choice was made on the basis that it is this model (Fig. 1) in which the three-stage concept of carcinogenesis is best exemplified, at least at our present state of knowledge.

1. The Moolgavkar-Knudson model and multistage hepatocarcinogenesis in the rat.

As described in Table 1, the three stages of initiation, promotion, and progression involve two genetic events separated by the epigenetic process of promotion. Similarly, in the Moolgavkar-Knudson model (Fig. 1), the first and second events are separated by the appearance of "intermediate cells." The mean number of intermediate (initiated) cells at time t is given by the formula

$$E(t) = \mu_1 \int_0^t X(s) \ \exp[(\alpha_2 - \beta_2)(t-s)]ds \qquad (1)$$

where

$X(s)$ = the number of normal susceptible cells at time (age) s

μ_1 = the rate of the "first event" (initiation)

α_2 = the rate of division of "intermediate cells" (cells in the stage of promotion)

β_2 = rates of "differentiation" and "death" of "intermediate cells" (apoptosis and/or differentiation of cells in the stage of promotion).

An approximate formula for the incidence function is given by $I(t) \sim \mu_2 E(t)$, where μ_2 is the rate of the "second event" (progression). This formula for $I(t)$ has been applied to human cancer incidence data. Unfortunately its use for the analysis of animal data is inappropriate because the probability of cancer is high (Moolgavkar et al., 1988). The exact incidence function, which is much more complicated, must be used. In particular, this function depends upon α_2 and β_2 individually, and not only upon their difference.

Each of the parameters of the model may be determined experimentally in multistage hepatocarcinogenesis in the rat, although at present the quantitative determination of μ_2 is less than satisfactory. For the purposes of this discussion, we will derive some of the above parameters from information in the literature and others from published and unpublished work in our own laboratory.

a.   Derivation of parameters for the model

Since there is already considerable knowledge of the normal growth, replicative rate, and turnover of hepatocytes in the rat, several of the parameters of equation (1) can be determined from the literature. Other parameters can be estimated with reasonable accuracy from one or another of the experimental protocols developed for the study of multistage hepatocarcinogenesis in rat liver. A protocol developed in our laboratory (Pitot, 1989) that closely mimics the model seen in Figure 1 is that of the initiation-promotion-initiation format first suggested by Potter (1981) and since utilized in analogous forms in models of multistage carcinogenesis in mouse skin (Hennings et al., 1985) and rat liver (Scherer, 1987). Rats are initiated during the first week of age with a low, noncarcinogenic dose of diethylnitrosamine (DEN); promotion is begun at weaning; a second initiation with a mitotic stimulus (70% partial hepatectomy) is performed at 4 months; and

the animals are sacrificed at 8 months. In this format
initiation with DEN is equivalent to the "first event". The
administration of a second agent, termed a "progressor"
agent, which is usually a direct-acting carcinogen (Pitot,
1989) that does not need metabolism to an ultimate reactive
form, represents the "second event" or the introduction
into the stage of progression.

From this format or the simpler initiation-promotion
protocol (Pitot et al., 1978), one may derive, from experimentation or from the literature, values for the parameters
of equation (1). With the initiation-promotion format, at
2 months of age the total number of hepatocytes in rat
liver can be estimated as 1.35×10^9 (Doljanski, 1960), and
about 40% of these cells are in DNA synthesis after a 70%
partial hepatectomy (Fabrikant, 1968). Pegg and Balog
(1979) have demonstrated that the $t_{1/2}$ time of maximal
ethylation of DNA by DEN in rat liver is 1 h. Therefore,
X_s in these animals at this time of life would be (1.35 x
10^9) x 0.4 x 0.3, or 162×10^6 hepatocytes. Since experimental studies in this laboratory (Pitot, 1989) have demonstrated that the number of AHF induced by the standard subcarcinogenic dose of DEN used in this format is approximately 20,000, μ_1 becomes 20,000 ÷ X_s or 1.25×10^{-4}.
Unfortunately, μ_2 cannot readily be determined at present,
although one might suggest that this parameter is equal to
the "promoter-independent" AHF (Hendrich et al., 1986) in
view of the reversibility of the stage of promotion and the
effects of promoting agents (see above).

Some of the parameters of the model can be estimated
from studies previously done in this laboratory by Rizvi et
al. (1988). In this study, the volume fraction of altered
hepatic foci was measured at different time points after
initiation with a standard subcarcinogenic dose of DEN and
70% partial hepatectomy followed by promotion with 0.05%
phenobarbital until sacrifice. Under the assumption that
the total number of hepatocytes is 1×10^9 (Doljanski,
1960), the mean number of cells in AHF can be computed from

the volume fraction (Table 2). Then, under the assumption that there is no spontaneous initiation, equation 1 implies that

$$\ln E(T) = \ln(\mu_1 X) + (\alpha_2 - \beta_2)t$$

where X is the number of susceptible hepatocytes at the time of initiation, and t now represents the length of time on promotion. Thus $\ln E(t)$ is a linear function of t with intercept $\ln(\mu_1 X)$ and slope $(\alpha_2 - \beta_2)$.

Table 2

Observed and expected number of cells in AHF with different promotion regimens. Observed numbers were calculated from volume fraction under the assumption that the total number of hepatocytes is 1×10^9. Expected numbers calculated from the model with estimated parameters $\mu_1 X = 1.13 \times 10^6$ and $\alpha_2 - \beta_2 = 6.10$ per cell per year. See text for details.

Months on 0.05% PB	Volume % fraction of AHF	Observed cell no. in AHF	Expected cell no. in AHF
2	0.265	0.265×10^7	0.312×10^7
5	1.83	1.83×10^7	1.43×10^7
8	6.949	6.95×10^7	6.59×10^7
10	16.067	16.1×10^7	18.20×10^7
12	18.971	18.9×10^7	50.00×10^7

When we plotted the logarithm of the observed number of cells in AHF against time, we found that the first four points (2, 5, 8, and 10 months) lay on a straight line. From these four points, the linear regression estimates of the parameters were $\mu_1 X = 1.13 \times 10^6$ and $(\alpha_2 - \beta_2) = 6.1$ per cell per year. Note the excellent agreement between observed and expected number of cells in AHF (Table 2 and Figure 2) for the first four time points. At 12 months of promotion, however, the model greatly overpredicts the number of cells, suggestive that the net growth rate of AHF decreases with prolonged promotion.

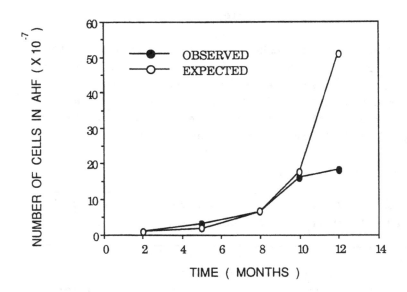

FIGURE 2. Observed and expected numbers of cells in AHF, where expectations are calculated from parameters above. These parameters were estimated from "best"-fit of model to the first four data points (i.e., 2, 5, 8, 10 months). Note the excellent fit to these points. There is no indication of a decrease in $(\alpha_2 - \beta_2)$ except at 12 months.

The parameter $(\alpha_2 - \beta_2)$ can be estimated from studies previously done in this laboratory by Rizvi et al. (1988).

In this study, changes in the volume percentage fraction of altered hepatic foci at 2, 5, and 8 months of feeding 0.05% phenobarbital as well as other data points were determined. We have utilized these data points to calculate the number of cells in the intermediate stage (that of promotion), using equation (1).

D. Implications and Conclusions from the Correlation of the Mathematical and Biological Models

Although a number of assumptions were required for the calculations seen above, it is apparent that with careful experimentation far more exact values for these parameters can be obtained from this biological model of multistage hepatocarcinogenesis. However, even with the calculations as shown, it is apparent that biological and mathematical models are quite compatible. On the basis of this relationship it then becomes possible to make distinct predictions from altering parameters in the mathematical model to what might be expected in the biological model. Such relationships should allow one to make predictions concerning the potency of specific promoting agents and, in the future, of complete carcinogens on the basis of dose-response relationships from the input of the mathematical model only. While it is not the intent of this report to carry through such exercises, the example given above illustrates the potential for such future relationships.

BIBLIOGRAPHY

M. ALAVANJA, J. ARON, C. BROWN and J. CHANDLER. Cancer risk-assessment models: anticipated contributions from biochemical epidemiology. J. Natl. Cancer Inst., 78, (1987), pp. 633-643.

B. N. AMES, L. S. GOLD, C. B. SAWYER and W. HAVENDER. Carcinogenic Potency. In Environmental Mutagens and Carcinogens. T. Sugimura, S. Kondo, and H. Takebe (eds.). University of Tokyo Press, Tokyo, (1982), pp. 663-670.

P. ARMITAGE and R. DOLL. The age distribution of cancer and a multi-stage theory of carcinogenesis. Brit. J. Cancer, 8, (1954), pp. 1-12.

R. K. BOUTWELL. The function and mechanism of promoters of carcinogenesis. CRC Crit. Rev. Toxicol., 2, (1974), pp. 419-443.

A. R. BUCKLEY, C. W. PUTNAM and D. H. RUSSELL. Prolactin is a tumor promoter in rat liver. Life Sci., 37, (1985), pp. 2569-2575.

W. BURSCH, B. LAUER, I. TIMMERMANN-TROSIENER, G. BARTHEL, J. SCHUPPLER and R. SCHULTE-HERMANN. Controlled death (apoptosis) of normal and putative preneoplastic cells in rat liver following withdrawal of tumor promoters. Carcinogenesis, 5, (1984), pp. 453-458.

K. C. CHU. A nonmathematical view of mathematical models for cancer. J. Chron. Dis., 40, (1987), pp. 163S-170S.

T. L. DAO and P.-C. CHAN. Hormones and dietary fat as promoters in mammary carcinogenesis. Environ. Health Persp., 50, (1983), pp. 219-225.

S. S. DEVESA, D. T. SILVERMAN, J. L. YOUNG, JR., E. S. POLLACK, C. C. BROWN, J. W. HORM, C. L. PERCY, M. H. MYERS, F. W. McKAY and J. F. FRAUMENI, JR. Cancer incidence and mortality trends among whites in the United States, 1947-84. J. Natl. Cancer Inst., 79, (1987), pp. 701-770.

L. DOCK, M. MARTINEZ and B. JERNSTRÖM. Induction of hepatic glutathione S-transferase activity by butylated hydroxyanisole and conjugation of benzo[a]pyrene diol-epoxide. Carcinogenesis, 5, (1984), pp. 841-844.

F. DOLJANSKI. The growth of the liver with special reference to mammals. Int. Rev. Cytol., 10, (1960), pp. 217-241.

H. E. DRIVER and A. E. M. McLEAN. Dose-response relationship for phenobarbitone promotion of liver tumours initiated by single dose dimethylnitrosamine. Br. J. Exp. Path., 67, (1986), pp. 131-139.

P. EMMELOT and E. SCHERER. The first relevant cell stage in rat liver carcinogenesis. A quantitative approach. Biochim. Biophys. Acta, 605, (1980), pp. 247-304.

J. I. FABRIKANT. The kinetics of cellular proliferation in regenerating liver. J. Cell Biol., 36, (1968), pp. 551-565.

E. FARBER and R. CAMERON. The sequential analysis of cancer development. Adv. Cancer Res., 31, (1980), pp. 125-226.

L. FOULDS. The experimental study of tumor progression: a review. Cancer Res., 14, (1954), pp. 327-339.

L. FOULDS. Multiple etiologic factors in neoplastic development. Cancer Res., 25, (1965), pp. 1339-1347.

D. W. GAYLOR and R. E. SHAPIRO. Extrapolation and risk estimation for carcinogenesis. *In* Advances in Modern Toxicology, Vol. 1, Part 2, New Concepts in Safety Evaluation, M. A. Mehlman, R. E. Shapiro, and H. Blumenthal, (eds.). John Wiley & Sons, New York, (1979).

T. GOLDSWORTHY, H. A. CAMPBELL and H. C. PITOT. The natural history and dose-response characteristics of enzyme-altered foci in rat liver following phenobarbital and diethylnitrosamine administration. Carcinogenesis, 5, (1984), pp. 67-71.

B. HANES and T. WEDEL. A selected review of risk models: one hit, multihit, multistage, probit, Weibull, and pharmacokinetic. J. Amer. Coll. Toxicol., 4, (1985), pp. 271-278.

M. HANIGAN and H. C. PITOT. Growth of carcinogen-altered rat hepatocytes in the liver of syngeneic recipients promoted with phenobarbital. Cancer Res., 45, (1985), pp. 6063-6070.

S. HENDRICH, H. P. GLAUERT and H. C. PITOT. The phenotypic stability of altered hepatic foci: effects of withdrawal and subsequent readministration of phenobarbital. Carcinogenesis, 7, (1986), pp. 2041-2045.

S. HENDRICH, H. P. GLAUERT and H. C. PITOT. Dietary effects on initiation and promotion of hepatocarcinogenesis in the rat. J. Cancer Res. Clin. Oncol., 114, (1988), pp. 149-157.

H. HENNINGS, R. SHORES, P. MITCHELL, E. F. SPANGLER and S. H. YUSPA. Induction of papillomas with a high probability of conversion to malignancy. Carcinogenesis, 6, (1985), pp. 1607-1610.

N. HIROTA and T. YOKOYAMA. Comparative study of abnormality in glycogen storing capacity and other histochemical phenotypic changes in carcinogen-induced hepatocellular preneoplastic lesions in rats. Acta Pathol. Jpn., 35(5), (1985), pp. 1163-1179.

T. ISHIKAWA, S. TAKAYAMA and T. KITAGAWA. Correlation between time of partial hepatectomy after a single treatment with diethylnitrosamine and induction of adenosinetriphosphatase-deficient islands in rat liver. Cancer Res., 40, (1980), pp. 4261-4264.

O. H. IVERSEN. TPA (12-O-tetradecanoyl-phorbol-13-acetate) as a carcinogen for mouse skin. Virchows Arch [Cell Pathol], 49, (1985), pp. 129-135.

T. KITAGAWA, R. WATANABE and H. SUGANO. Induction of γ-glutamyl transpeptidase activity by dietary phenobarbital in "spontaneous" hepatic tumors of C3H mice. Gann, 71, (1980), pp. 536-542.

R. J. KOCIBA, D. G. KEYS, J. E. BEYER, R. M. CARREON, C. E. WADE, D. A. DITTENBER, R. P. KALNINS, L. E. FRAUSON, C. N. PARK, S. D. BARNARD, R. A. HUMMEL and C. G. HUMISTON. Results of a two-year chronic toxicity and oncogenicity study of 2,3,7,8-tetrachlorodibenzo-*p*-dioxin in rats. Toxicol. Appl. Pharmacol., 46, (1978), pp. 279-303.

Y. KURATA, H. TSUDA, S. TAMANO and N. ITO. Inhibitory potential of acetaminophen and *o*-, *m*-, *p*-aminophenols for development of γ-glutamyltranspeptidase-positive liver cell foci in rats pretreated with diethylnitrosamine. Cancer Lett., 28, (1985), pp. 19-25.

G.-H. LEE, N. SAWADA, Y. MOCHIZUKI, K. NOMURA and T. KITAGAWA. Immortal epithelial cells of normal C3H mouse liver in culture: possible precursor populations for spontaneous hepatocellular carcinoma. Cancer Res., 49, (1989), pp. 403-409.

S. H. MOOLGAVKAR. Carcinogenesis modeling: from molecular biology to epidemiology. Annu. Rev. Public Health, 7, (1986), pp. 151-169.

S. H. MOOLGAVKAR, A. DEWANGI AND D. J. VENZON. A stochastic two-stage model for cancer risk assessment. I. The hazard function and the probability of tumor. Risk Anal., 8, (1988), pp. 383-392.

S. H. MOOLGAVKAR and A. G. KNUDSON, JR. Mutation and cancer: A model for human carcinogenesis. J. Natl. Cancer Inst., 66, (1981), pp. 1037-1052.

M. A. MOORE, K. NAKAGAWA, K. SATOH, T. ISHIKAWA and K. SATO. Single GST-P positive liver cells - putative initiated hepatocytes. Carcinogenesis, 8, (1987), pp. 483-486.

H. J. MULLER. Radiation damage to the genetic material. Sci. in Prog., 7, (1951), p. 130.

C. O. NORDLING. A new theory on the cancer inducing mechanism. Br. J. Cancer, 7, (1953), pp. 68-72.

P. C. NOWELL and H. P. MORRIS. Chromosomes of "minimal deviation" hepatomas: a further report on diploid tumors. Cancer Res., 29, (1969), pp. 969-970.

K. OGAWA, T. ONOE and M. TAKEUCHI. Spontaneous occurrence of gamma-glutamyl transpeptidase-positive hepatocytic foci in 105-week-old Wistar and 72-week-old Fischer 344 male rats. J. Natl. Cancer Inst., 67, (1981), pp. 407-412.

A. E. PEGG and B. BALOG. Formation and subsequent excision of O^6-ethylguanine from DNA of rat liver following administration of diethylnitrosamine. Cancer Res., 39, (1979), pp. 5003-5009.

H. C. PITOT. Drugs as promoters of carcinogenesis. The Induction of Drug Metabolism Symposium, Ashford Castle, Ireland, May 24-27, 1978. R. W. Estabrook and

E. Lindenlaub, (eds.), p. 471. FK Schattauer Verlag, Stuttgart-New York, (1979).

H. C. PITOT. Characterization of the stage of progression in hepatocarcinogenesis in the rat. *In* Boundaries between Promotion and Progression, O. Sudilovsky, L. Liotta, and H. C. Pitot (Eds.), Plenum Press, 1989.

H. C. PITOT, L. BARSNESS, T. GOLDSWORTHY, et al. Biochemical characterization of stages of hepatocarcinogenesis after a single dose of diethylnitrosamine. Nature (Lond.), 271, (1978), pp. 456-458.

H. C. PITOT, L. E. GROSSO and T. GOLDSWORTHY. Genetics and epigenetics of neoplasia: facts and theories. Carcinogenesis, 10, (1985), pp. 65-79.

H. C. PITOT, T. L. GOLDSWORTHY, S. MORAN, W. KENNAN, H. P. GLAUERT, R. R. MARONPOT and H. A. CAMPBELL. A method to quantitate the relative initiating and promoting potencies of hepatocarcinogenic agents in their dose-response relationships to altered hepatic foci. Carcinogenesis, 8, (1987), pp. 1491-1499.

H. C. PITOT, D. BEER and S. HENDRICH. Multistage carcinogenesis: the phenomenon underlying the theories. *In* Theories of Carcinogenesis, O. Iversen, (ed.), pp. 159-177. Hemisphere Press, (1988).

M. POLLARD, P. H. LUCKERT and R. A. ADAMS. Detection of gamma glutamyl transpeptidase in the livers of germ-free and conventional rats. Lab. Animal Sci., 32, (1982), pp. 147-149.

J. A. POPP, B. H. SCORTICHINI and L. K. GARVEY. Quantitative evaluation of hepatic foci of cellular alteration occurring spontaneously in Fischer-344 rats. Fund. Appl. Toxicol., 5, (1985), pp. 314-319.

V. R. POTTER. A new protocol and its rationale for the study of initiation and promotion of carcinogens in rat liver. Carcinogenesis, 2, (1981), pp. 1375-1379.

T. A. RIZVI, W. KENNAN, Y.-H. XU and H. C. PITOT. The effects of dose and duration of administration on the promotion index of phenobarbital in multistage hepatocarcinogenesis in the rat. Acta Pathol. Microbiol. Immunol., 96, (1988), pp. 262-268.

L. ROSSI, M. RAVERA, G. REPETTI and L. SANTI. Long-term administration of DDT or phenobarbital-Na in Wistar rats. Int. J. Cancer, 19, (1977), pp. 179-185.

L. SARGENT, Y.-H. XU, G. L. SATTLER, L. MEISNER and H. C. PITOT. Ploidy and karyotype of hepatocytes isolated from enzyme-altered foci in two different protocols of multistage hepatocarcinogenesis in the rat. Carcinogenesis, 10, (1989), pp. 387-391.

E. SCHERER. Neoplastic progression in experimental hepatocarcinogenesis. Biochim. Biophys. Acta, 738, (1984), pp. 219-236.

E. SCHERER. Relationship among histochemically distinguishable early lesions in multistep-multistage hepatocarcinogenesis. Arch. Toxicol. Suppl., 10, (1987), pp. 81-94.

R. SCHULTE-HERMANN. Tumor promotion in the liver. Arch. Toxicol., 57, (1985), pp. 147-158.

R. SCHULTE-HERMANN, I. TIMMERMANN-TROSIENER and J. SCHUPPLER. Promotion of spontaneous preneoplastic cells in rat liver as a possible explanation of tumor production by nonmutagenic compounds. Cancer Res., 43, (1983), pp. 839-844.

J. D. SCRIBNER and R. SÜSS. Tumor initiation and promotion. Int. Rev. Exp. Pathol., 18, (1978), pp. 137-197.

A. E. SIRICA, J. K. JICINSKY and E. K. HEYER. Effect of chronic phenobarbital administration on the gamma-glutamyl transpeptidase activity of hyperplastic liver lesions induced in rats by the Solt/Farber initiation: selection process of hepatocarcinogenesis. Carcinogenesis, 5, (1984), pp. 1737-1740.

T. J. SLAGA. Overview of tumor promotion in animals. Environ. Health Persp., 50, (1983), pp. 3-14.

J. A. SWENBERG, F. L. RICHARDSON, J. A. BOUCHERON, F. H. DEAL, S. A. BELINSKY, M. CHARBONNEAU and B. G. SHORT. High to low-dose extrapolation; critical determinants involved in the dose response of carcinogenic substances. Environ. Health Perspect., 76, (1987), pp. 57-63.

H. S. TAPER. The effect of estradiol-17-phenylpropionate and estradiol benzoate on N-nitrosomorpholine-induced liver carcinogenesis in ovariectomized female rats. Cancer, 42, (1978), pp. 462-467.

H. TSUDA, T. MASUI, K. IMAIDA, S. FUKUSHIMA and N. ITO. Promotive effect of primary and secondary bile acids on the induction of γ-glutamyl transpeptidase-positive liver cell foci as a possible endogenous factor for hepatocarcinogenesis in rats. Gann, 75, (1984), pp. 871-875.

H. URA, A. DENDA, Y. YOKOSE, M. TSUTSUMI and Y. KONISHI. Effect of vitamin E on the induction and evolution of enzyme-altered foci in the liver of rats treated with diethylnitrosamine. Carcinogenesis, 8, (1987), pp. 1595-1600.

A. K. VERMA and R. K. BOUTWELL. Effects of dose and duration of treatment with the tumor-promoting agent, 12-O-tetradecanoylphorbol-13-acetate on mouse skin carcinogenesis. Carcinogenesis, 1, (1980), pp. 271-276.

I. R. WANLESS and A. MEDLINE. Role of estrogens as promoters of hepatic neoplasia. Lab. Invest., 46, (1982), pp. 313-320.

J. M. WARD. Background data and variations in tumor rates of control rats and mice. Progr. Exp. Tumor Res., 26, (1983), pp. 241-258.

G. P. WARWICK. Effect of the cell cycle on carcinogenesis. Fed. Proc., 30, (1971), pp. 1760-1765.

L. W. WATTENBERG. Inhibition of chemical carcinogenesis. J. Natl. Cancer Inst., 60, (1978), pp. 11-18.

A. S. WHITTEMORE. Quantitative theories of oncogenesis. Adv. Cancer Res., 27, (1978), pp. 55-88.

G. M. WILLIAMS. Modulation of chemical carcinogenesis by xenobiotics. Fund. Appl. Toxicol., 4, (1984), pp. 325-344.

T. S. YING, K. ENOMOTO, D. S. R. SARMA and E. FARBER. Effects of delays in the cell cycle on the induction of preneoplastic and neoplastic lesions in rat liver by 1,2-dimethylhydrazine. Cancer Res., 42, (1982), pp. 876-880.

Cell Proliferation and Hepatocarcinogenesis

Michael Schwarz[1], Albrecht Buchmann[1], Larry W. Robertson[2]
and Werner Kunz[1]

Abstract. Chemically-induced hepatocarcinogenesis in
rodents is a very useful tool to study qualitatively and
quantitatively critical changes occurring during the
carcinogenic process. There is strong evidence to suggest
that enzyme-altered foci in liver are precursor lesions
causally related to malignant transformation. In
experiments with continuous exposure of rats to different
doses of hepatocarcinogens the existence of quantitative
relationships between the development of enzyme-altered
foci and subsequent tumor manifestation in liver was
demonstrated. The analysis of multiple enzyme markers
indicated a marked heterogeneity of phenotypes between
individual foci. Such diversity was also observed with
respect to the growth properties of individual lesions and
relationships between foci phenotype and proliferation
rate were demonstrated. Our results suggest that not only
the total number of enzyme-altered cells in liver but also
the proliferation rate of individual cell clones is of
major relevance for the transition(s) leading to malignant
cell populations. In initiation-promotion experiments
quantitative relationships between promoting activity of
various xenobiotics and their potency to induce adaptive
liver growth were established. Induction of cell
proliferation in normal liver and in preneoplastic liver
cells is therefore assumed to play an important role
during the carcinogenic process.

---

[1]German Cancer Research Center, Institute of Biochemistry,
Im Neuenheimer Feld 280, 6900 Heidelberg, FRG.

[2]University of Kentucky, Graduate Center for Toxicology,
Lexington, Ky, U.S.A

Abbreviations: ATPase: canalicular adenosine
triphosphatase; cyt.P-450: cytochrome P-450; DEN:
diethylnitrosamine (N-nitrosodiethylamine); 4-DAB: N,N-4-
dimethylaminoazobenzene; NDEOL: N-nitrosodiethanolamine;
NNM: N-nitrosomorpholine; EAF: enzyme-altered foci; p.h.:
partial hepatectomy.

I. Cell Proliferation and Initiation

It is well established from a large variety of studies
that cell proliferation, or more precisely DNA synthesis
in target cells, plays a very important role during tumor
initiation. Cells which undergo DNA synthesis seem to be
much more vulnerable to initiation by genotoxic
carcinogens than resting cells and it is possible that
only cells able to undergo DNA synthesis can be initiated.
The effects of cell proliferation on initiation of the
carcinogenic process have been extensively studied in the
regenerating rodent liver where a wave of synchronously
occurring DNA synthesis within hepatocytes can be induced
by surgical removal of two-thirds of the liver tissue.
Maximal increases in preneoplastic and neoplastic
responses in liver have been observed when initiating
carcinogens were administered during S-phase of
regenerative liver growth 20 to 24 hours after partial
hepatectomy (4,5,12,23,31,37) and it is generally assumed
that this effect is mediated by elevated mutation rates in
liver cells driven into DNA synthesis before they had time
to repair critical lesions in their genetic material. In
addition, partial hepatectomy (p.h.) has been found to be
effective when performed within a short time window after
carcinogen treatment. Two recent papers (8,15) describe in
detail the effect of varying the time period between DEN
application and p.h. using enzyme-altered (ATPase-
deficient) foci in rat liver as parameter of carcinogenic
effectiveness. Data from both groups indicate a strong
stimulatory effect of p.h. when the surgical manipulation
was performed within one to two days after a single
application of the hepatocarcinogen diethylnitrosamine
(DEN). Prolongation of the time period between carcinogen
treatment and p.h. resulted in a gradual decrease in
carcinogenic efficacy and no significant difference in
carcinogenic response was seen between animals treated
with DEN only and animals that underwent an additional
p.h. when the time lag between the treatments exceeded 15

days. Similarly, no effect on liver tumor incidence was observed in rats that underwent a p.h. two days after stop of continuous administration of 5 mg/kg diethylnitrosamine for 82 days as compared to rats treated with the carcinogen alone (29). This evidence demonstrates that p.h. alone, when performed long enough after carcinogen treatment, does not possess promoting activity. Since the removal of two-thirds of liver tissue during p.h. is associated with the removal of (approximately) two-thirds of initiated hepatocytes present in liver at the time of surgery, it has to be assumed that the remaining initiated hepatocytes respond to the growth stimulus as do their normal counterparts so that the proportion of initiated cells after completion of liver regeneration is similar to the one before liver surgery.

The ineffectiveness of p.h. when performed long enough after carcinogen treatment may be explained by the fact that at the later time points critical DNA-bound metabolites have been entirely eliminated by repair processes so that stimulation of DNA synthesis does not further increase the mutation rate within target cells. The half life of this hypothetical DNA adduct(s) has been calculated by Emmelot and Scherer (8) on the basis of their data to be in the range of 2 to 4 days.

The time-dependent decrease in effectiveness of stimulation of hepatocyte proliferation by p.h. suggests that these cells are only at risk to become initiated if they undergo DNA synthesis during a critical time window after carcinogen treatment. The same time window should in principle be valid for cells that proliferate in "normal liver", without stimulation by p.h. This might be used to get a rough estimate on the number of liver cells at risk to become initiated in experiments which include treatment with DEN as an initiating agent.

Adult rat liver contains approximately $9 * 10^8$ hepatocytes (taken from reference 38, assuming a liver weight of 7 g and 30% of hepatocytes being binucleated). About 2.5 million hepatocytes per day enter DNA synthesis

(calculated from reference 26, which gives a 42 hours-[3]H-
thymidine labeling index of 0.48%). Let us assume that
initiated cells stem from cells which undergo DNA
synthesis before repair of a critical adduct in their DNA
has been completed and that DNA synthesis has to occur
within a 15-day time window after DEN treatment. The
number of hepatocytes which enter DNA synthesis within
this time period amounts to about 37.5 million cells. Out
of this population of cells 200 to 300 hepatocytes per
liver per mg DEN/kg body wt. (without p.h.) become
clonogens of ATPase-deficient foci (this number is taken
from reference 8 and is in accordance with own unpublished
observations; for definition of clonogens see K. Clifton,
this volume).

II. Enzyme-altered Foci and Liver Tumors

Shortly following exposure of experimental animals to
hepatocarcinogens foci of cells appear in liver that are
characterized by changes in the expression of several
marker enzymes which can be used for their identification
(1,2,3,10,11,13,16,19,23,24; see also Pitot et al., this
volume). These focal lesions are monoclonal in origin
(27,40) and display in comparison to their surrounding
normal hepatocytes increases in cell proliferation
(26,28,30,31). There are several lines of evidence to
suggest that these enzyme-altered foci (EAF) in liver are
precursors for tumors in this organ. A variety of marker
enzyme changes observed within EAF are similar or
identical to the ones observed within liver tumors. Focal
clones exhibiting more progressed neoplastic
characteristics (foci within foci) can - especially under
certain treatment protocols - be observed within existing
enzyme-altered lesions (32). In addition, we and others
have demonstrated the existence of quantitative
relationships between the total volume of enzyme-altered
tissue in liver and the subsequent development of liver

tumors (8,16,17,18). For this purpose we have used an
approach similar to the one described by Druckrey (6,7).

In his early studies Druckrey demonstrated for a large
variety of carcinogens the existence of quantitative
relationships between exposure level and the median times
of tumour induction yielding stright lines when plotted in
a double-logarithmic plot with slopes being characteristic
for the carcinogen under investigation. In an experiment
with chronic treatment of rats with the hepatocarcinogen
N-nitrosomorpholine (NNM) we have determined this
relationship for the induction of liver tumors by this
carcinogen (see Figure 1).

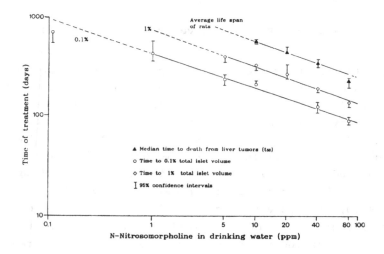

Figure 1: Dose-time relationships for the induction of
liver tumors and defined fractions of ATPase-deficient
foci in rat liver by N-nitrosomorpholine (NNM). Groups of
rats were treated continuously with different doses of the
hepatocarcinogen. Animals were sacrificed sequentially and
the time periods necessary to reach defined fractions of
enzyme-altered tissue in liver (0.1% or 1%, respectively)
were calculated for each dose group by regression analysis
of the time-dependent increases in the volumetric fraction
of ATPase-deficient foci. Additional animals were used to
determine median times to death from liver tumors. (Figure
modified from Kunz et al., 1983).

In this experiment, additional animals were killed at earlier time points and the evolution of ATPase-deficient foci in liver was determined. From these data we then calculated the time periods necessary to reach an arbitrarily chosen end point (in this case 0.1 or 1 % of liver being occupied by ATPase-deficient tissue). Those time points were then plotted in the same double-logarithmic net against dose of NNM. The slopes of these regression lines were found to be almost identical to the one observed for liver tumor induction (Figure 1). This enables to study the carcinogenic effect of this nitrosamine not only much earlier, but also at those low dose levels were the development of tumors would exceed the average life span of the animals (17). Additional evaluations performed on the basis of this data set will be reported in this issue (see Moolgavkar et al.).

We have then focussed our interest on two additional hepatocarcinogens, namely 4-dimethylaminoazobenzene (4-DAB, also known as butter yellow) and N-nitroso-diethanolamine (NDEOL). The carcinogenic potencies of 4-DAB and NDEOL have been studied in detail by Druckrey and coworkers (6,7) and by Preussmann et al. (25), respectively.

The increases in the volumetric fraction occupied by ATPase-deficient tissue in liver of rats treated continuously with either 4-DAB or NDEOL were quantitated and the time points where 0.1 or 0.5% volumetric fraction were reached were calculated from the regresssion lines for each dose group. In analogy to the NNM-experiment described above these time points were then plotted against dose in a double-logarithmic net (see Figure 2).

The slopes of the regression lines describing the preneoplastic responses were found to be similar to the ones observed with respect to tumor formation. The dose-dependency of islet and tumor induction, however, was considerably different for 4-DAB and NDEOL. Within the given dose-range, doubling the daily exposure level of 4-DAB will cut the median tumor induction time into half

whereas with NDEOL an approximately 50-fold increase in
daily dose level is necessary to obtain the same effect.
As a consequence 4-DAB is a very potent carcinogen at a
high dose level resulting in short tumor induction times
but becomes a weak carcinogen when lowering the dose level
by just one order of magnitude (see Figure 2). With NDEOL
tumor induction times are much longer, even at very high
dose levels, but the carcinogenic response is to a much
lesser extent affected by changes in dose.

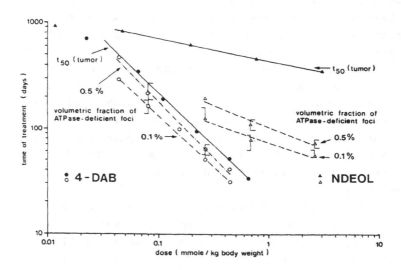

Figure 2: Dose-time relationships for the induction of
liver tumors and defined fractions of ATPase-deficient
foci in rat liver by 4-dimethylaminoazobenzene (4-DAB and
N-nitrosodiethanolamine (NDEOL). Liver tumor data are
taken from the literature (4-DAB: Druckrey, 1967; NDEOL:
Preussmann et al., 1982). For the analysis of
preneoplastic response rats were treated continuously with
different doses of 4-DAB or NDEOL. For further details see
legend to Figure 1.

A second very characteristic difference between the two carcinogens became obvious when analyzing the time periods between the induction of early enzyme-altered foci and liver tumours. These time periods are very long for NDEOL but much shorter for 4-DAB. From this we concluded that 4-DAB may possess a dose-related potency to accelerate the process from the early enzyme-altered foci to the liver tumors, an activity which we referred to as "intrinsic promoting" activity (34). NDEOL seems to lack such an activity. However, foci initiated by NDEOL treatment can be effectively promoted by subsequent 4-DAB administration (34).

Promotion of the carcinogenic process by 4-DAB is assumed to be mediated by a strong stimulatory effectiveness of this compound on the growth of enzyme-altered foci. The difference between 4-DAB and NDEOL in their potency to enhance the proliferation of enzyme-altered lesions in liver can be indirectly taken from data shown in Figure 3. At two selected dose-levels of 4-DAB (0.06% in diet) and NDEOL (0.2% in drinking water) there was an almost identical increase in the volumetric fraction of ATPase-deficient tissue in liver when plotted as a function of time of treatment with these carcinogens. Therefore, with respect to this parameter these dose levels of 4-DAB and NDEOL were almost equipotent. Upon treatment with NDEOL, however, the increase in the volumetric fraction of EAF was predominantly mediated by an enhancement of foci number - indicating comparatively strong initiating potency of this carcinogen, whereas with 4-DAB it was mostly due to an enhancement of individual foci size. Consequently, the mean islet volume of foci generated by continuous 4-DAB treatment was always higher than that of foci induced by NDEOL (approximately 10-fold increase at 100 days of treatment; see also Figure 3).

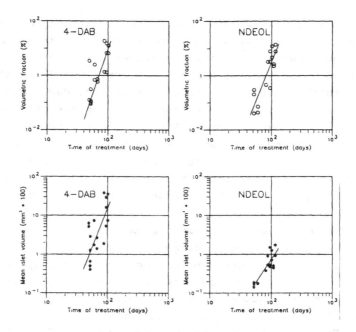

Figure 3: Development of ATPase-deficient foci in livers of rats treated continuouly with 4-DAB and NDEOL. 4-DAB was given in diet at a concentration of 0.06% and NDEOL was given in drinking water at a concentration of 0.2%. Rats were sacrificed sequentially, sections were taken from frozen liver blocks and stained for ATPase activity. Upper two graphs: Relative fraction of enzyme-altered tissue in liver. Lower two graphs: Mean islet volume. Each point represents a value from one animal; regression lines are also given to facilitate comparison of data (taken from Schwarz et al., 1984).

NDEOL and 4-DAB thus exhibit remarkably different effects on the development of enzyme-altered foci in liver. At dose levels leading to an approximately equivalent mass of enzyme-altered cells, NDEOL produces numerous but small liver foci, whereas continuous treatment with 4-DAB leads to the appearance of only few, but rapidly growing liver lesions. Since the time periods between the induction of the early enzyme-altered lesions and the induction of liver tumors were very short with 4-DAB and much longer with NDEOL, it can be concluded that

the probability of changes leading from the tumor
progenitor cells to the tumor cells is somehow related to
the growth properties of the individual preneoplastic cell
clones and not simply to the total number of enzyme-
altered intermediate cells. This fact is of importance for
carcinogenicity models that consider cell birth and death
of intermediate cell populations as important factors for
the velocity of the carcinogenic process (20,21; see also
several contributions to this topic in this issue).

III. Phenotype and Growth Behaviour of Enzyme-altered Foci

The differences in the growth rate of enzyme-altered
foci are reflected by a marked heterogeneity of foci
phenotypes. The relationship between phenotype and
proliferation kinetics was initially studied in stop
experiments with diethylnitrosamine (DEN) as model
carcinogen. Discrimination into different phenotypes was
performed on the basis of serial liver sections stained
for a variety of different marker enzymes such as
canalicular ATPase, g-glutamyltranspeptidase (GGT),
epoxide hydrolase (EH) and various cytochrome P-450 (cyt.
P-450) isozymes. Characteristic changes in the expression
of these enzymes during the process of
hepatocarcinogenesis were observed which are reported
elsewhere in more detail (2,3,19). In these studies
"early" enzyme-altered foci appearing shortly after stop
of carcinogen treatment showed unchanged or even slightly
increased cyt. P-450 levels. In the subsequent time course
of carcinogenesis various cyt. P-450 isozymes started to
decline and expansively growing neoplastic nodules were
generally characterized by decreases in all four cyt. P-
450 isozymes investigated.

Quantitative data on number and size of lesions of two
selected phenotypes observed at various time points after
stop of DEN treatment are given in Figure 4. The number of
foci showing ATPase-deficiency without any additional
marker enzyme change (ATPase only) was comparatively high

and did not change significantly throughout the experiment. In contrast, the number of lesions characterized by a decrease in three or four cyt. P-450 isozymes along with the ATPase change was only very small. The volumetric fraction in liver occupied by lesions of this latter phenotype, however, showed a dramatic increase with progression of time thus indicating a strong growth advantage over foci with ATPase change only.

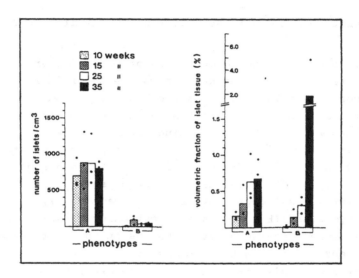

Figure 4: Development of different foci phenotypes in rat liver following limited DEN exposure. DEN (50 or 100 ppm) was given in drinking water for 10 days. Carcinogen-induced foci were subdivided into two phenotypes: A; lesions showing a decrease in the marker enzyme ATPase without additional changes in any out of four cytochrome P-450 isozymes investigated. B; lesions showing ATPase deficiency plus a decrease in three or four cyt. P-450 isozymes. Points: values from one animal. Bars: Values from three animals. Foci numbers of phenotype B were significantly lower than that of phenotype A (p = 0.002). The volumetric fraction in liver of lesions of phenotype A increased about linearly with time. In contrast, volume fractions of lesions of phenotype B showed an increase with about the fourth power of time as was found by linear regression analysis of double-logarithmically transformed data (p = 0.003). (Data taken from Buchmann et al., 1987).

A variation in the frequency of lesions of the various
phenotypes was also observed in experiments with
continuous exposure of rats to 4-DAB and NDEOL. With
NDEOL, a comparatively high number of foci exhibited only
ATPase deficiency without additional changes in the other
markers investigated. Moreover, only a small number of
NDEOL induced lesions showed concomitant alterations in
all markers. In contrast, continuous 4-DAB treatment led
to a comparatively large number of lesions exhibiting
changes in the expression of ATPase along with a decrease
in the level of four cytochrome P-450 isozymes (35,36).
These findings demonstrate a relationship between
phenotype and growth of lesions produced under various
conditions.

Although it is not clear at present whether or not the
enzyme alterations observed in preneoplastic and
neoplastic cells are causally related to the process of
malignant transformation, the analysis of enzyme
alterations appears to be a good tool with which to
discriminate and characterize islet subpopulations with
different neoplastic potential and fate. A link between
the expression of drug-metabolizing enzymes and the
proliferative potential of preneoplastic cells has been
drawn by Farber and his collegues (9). These authors
suggested that a decrease in the cytochrome P-450 system
will allow preneoplastic cells to escape from toxic
environmental effects, thus leading to a selective
proliferation of these cells. This hypothesis gives a
reasonable explanation for the extremely rapid growth of
nodular lesions observed in experimental systems
associated with chronic toxicity as is present during
treatment with high doses of 4-DAB. Liver toxicity from
this carcinogen is believed to result from intermediates
of metabolic activation via cyt. P-450. The decrease in
activity of this enzyme system will protect prenoplastic
liver cells against toxicity, whereas normal liver cells -
expressing high cytochrome P-450 levels - will form toxic
metabolites which lead to an inhibition of proliferation

of these cells. Besides this mechanism of selective
resistance to cytotoxic effects, however, alterations in
the endogenous regulation of cell proliferation may also
be operative, leading to autonomeous growth of
preneoplastic cells. The selective outgrowth of cyt. P-450
deficient lesions during hepatocarcinogenesis would then
be related to alterations in regulatory systems which
affect both the expression of the cyt. P-450 system and
the growth-controlling principles of the preneoplastic
cell population. In this case, the decrease in cyt. P-450
isozyme levels would only be a phenotypic indication for
alterations in the regulation of homeostasis within
preneoplastic and neoplastic cells.

IV. Effect of Tumor Promotors

It is known from a variety of studies that the
regulation of cell homeostasis can be modulated by
application of enzyme-inducing xenobiotics. In an
experiment including a series of different chemicals such
as phenobarbital (Pb), 3-methylcholanthrene (3-MC) and
various polychlorinated biphenyl congeners we have
investigated the promoting potency of these compounds in
liver along with certain cellular and biochemical response
parameters. In normal liver, administration of these
xenobiotics leads to induction of adaptive liver growth,
an effect which is mediated either by increases in
hepatocyte number or by polyploidisation of these cells or
by both mechanisms. These adaptive responses are fully
reversible upon inducer withdrawal.

Interestingly, the promoting activity of the various
xenobiotics investigated in this study was found to
correlate well with their potency to induce liver growth
(see Figure 5). Similar observations have been reported by
Nims et al. (22), comparing the liver growth inducing and
promoting properties of a series of different
barbiturates. A relationship between induction of liver
growth and promoting activity is further substantiated by

similarities in the dose-response curves for liver tumor
promotion and induction of organ growth with certain
polychlorinated biphenyls (14).

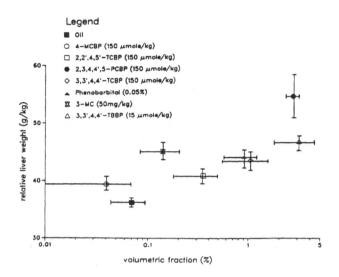

Figure 5: Comparison between the promoting activity of
various xenobiotics in liver and their potency to induce
adaptive liver growth. All rats were initiated with DEN
(5mg/kg on 10 consecutive days) and were subsequently
given the compounds indicated for a total of 8 weeks. PB
was given continuously in diet. All other compounds were
injected i.p. once a week at the dose levels indicated in
the Figure. Rats were killed 9 weeks after stop of
promoter treatment. Serial liver sections were stained for
ATPase and GGT activity. Preneoplastic response is
expressed as percentage of liver tissue occupied by foci
with decreased ATPase and coincident increase in GGT
activity. Induction of liver growth was determined in
animals treated for one week only. Values are expressed as
relative liver weight (g/kg body wt.). Data represent
means +/- S.D. Abbreviations are as follows: 4-MCBP, 4-
chlorobiphenyle; 2,2',4,5-TCBP, 2,2',4,5-tetrachloro-
biphenyle; 2,3,4,4',5-PCBP, 2,3,4,4',5-pentachloro-
biphenyle; 3,3',4,4'-TCBP, 3,3',4,4'-tetrachlorobiphenyle;
3,3',4,4'-tetrabromobiphenyle; 3-MC, 3-methylcholanthrene.
Details of the experiment will be reported elsewhere
(manuscript in preparation).

Since many of these liver tumor promoters lack overt liver toxicity, and in addition to this, induce the liver monooxygenase system in normal and in preneoplastic liver tissue, mechanisms of growth stimulation of intermediate preneoplastic cells via selective resistance to cytotoxicity cannot be brought into play to explain their promoting activity.

Alternative mechanisms have been reviewed in a recent paper by Schulte-Hermann (33). According to the hypothesis of this author initiated cells are committed to over-express an adaptive gene program expressed in normal liver to cope with the increased functional demands imposed by growth-stimulating tumor promotors. "Compounds such as phenobarbital would promote by providing the signal that trigger expression of this specific, but incompletely controlled program". Alternatively, or in addition to this, tumor promotors could act by blocking cell to cell communication, thus inhibiting the transfer of signals for growth restriction from normal hepatocytes to initiated cells (see Yamasaki and Fitzgerald, this issue).

The stimulation of proliferation of tumor precursor cells in liver is one of the prerequisites for promoting activity. Knowledge about the underlaying mechanisms of action are of importance for risk assessment of promoting agents since the promoting activity of different xenobiotics showing equipotent activity at high doses may differ considerably at the low dose range depending on whether stimulation of proliferation of the initiated cells by these compounds is mediated by selective resistance to cytotoxicity or by direct growth-stimulatory effects.

Acknowledgements:

The authors wish to express their thanks to Dr. R. Wolf and Dr. F. Oesch for gift of cytochrome P-450 specific antibodies. The excellent technical assistence of R. Schmitt and G. Robinson is greatly acknowledged.

Literature:

1.) **Bannasch, P., Hacker, H.J., Klimek, F., Mayer, D.:**
Hepatocellular glycogenesis and related pattern of
enzymatic changes during hepatocarcinogenesis.
in: Advances in Enzyme Regulation, Vol 22, G.Weber (Ed.),
Pergamon Press, Oxford, 97-121 (1984).

2.) **Buchmann, A., Kuhlmann, W., Schwarz, M., Kunz, W.,
Wolf, C.R., Moll, E., Friedberg, T., Oesch, F.:**
Regulation and expression of four cytochrome P450-
isoenzymes, NADPH-cytochrome P-450 reductase, the
glutathione transferases B and C and microsomal epoxide
hydrolase in preneoplastic and neoplastic lesions in rat
liver.
Carcinogenesis 6, 513-521 (1985).

3.) **Buchmann, A., Schwarz, M., Schmitt, R., Wolf, C.R.,
Oesch, F., Kunz, W.:**
Development of cytochrome P450 altered preneoplastic and
neoplastic lesions during nitrosamine-induced
hepatocarcinogenesis in the rat.
Cancer Res. 47, 2911-2918 (1987).

4.) **Columbano, A. Rajalakshmi, S., Sarma, D.S.R.:**
Requirement of cell proliferation for the initiation of
liver carcinogenesis as assayed by three different
procedures.
Cancer Res. 41, 2079-2083 (1981).

5.) **Craddock, V.:**
Liver carcinomas induced in rats by single administration
of dimethylnitrosamine after partial hepatectomy.
J. Natl. Cancer Inst. 47, 899-907 (1971).

6.) **Druckrey, H., Kupfmüller, K.:**
Quantitative Analyse der Krebsentstehung.
Zeitschrift für Naturforschung. 3b, 254-266 (1948).

7.) **Druckrey, H.:**
Quantitative aspects in chemical carcinogensis.
U.I.C.C. Monographs. 7, 60-78 (1967).

8.) **Emmelot, P., Scherer, E.:**
The first relevant cell stage in rat liver carcinogenesis:
a quantitative approach.
Biochim. Biophys. Acta 605, 247-304 (1980).

9.) **Farber, E., Cameron, R.:**
The sequential analysis of cancer development.
Adv. Cancer Res. 31, 125-226 (1980).

10.) **Friedrich-Freksa, H., Gössner, W., Börner, P.:**
Histochemische Untersuchungen der Cancerogenese in der
Rattenleber nach Dauergabe von Diäthylnitrosamin.
Z. Krebsforsch. 72, 226-239 (1969a).

11.) Friedrich-Freksa, H., Papadopulu, G., Gössner, W.:
Histochemische Untersuchungen der Cancerogenese in der
Rattenleber nach zeitlich begrenzter Verabfolgung von
Diäthylnitrosamin.
Z. Krebsforsch. 72, 240-253 (1969b).

12.) Glinos, A.D., Bucher, N.L.R., Aub, J.C:
The effect of liver regeneration on tumour formation in
rats fed 4-dimethylaminoazobenzene.
J.Exp.Med. 93, 313-324 (1951).

13.) Goldfarb, S., Pugh, T.D.:
Enzyme histochemical phenotypes in primary hepatocellular
carcinomas.
Cancer Res. 41, 2092-2095 (1981).

14.) Greim, H., Deml, E., Oesterle, D.:
Dose dependence and risk evaluation of the tumor-promoting
effects of phenobarbital and polychlorinated biphenyls in
hepatocarcinogenesis. In: Tumorpromotoren, Erkennung,
Wirkungsmechanismen und Bedeutung. BGA-Schriften 6: Appel,
K.E. und Hildebrandt, A.G. (Eds.),
MMV Medizin Verlag München, 76-94 (1985).

15.) Ishikawa, T., Takayama, S., Kitagawa, T.:
Correlation between the time of partial hepatectomy after
a single treatment with diethylnitrosamine and induction
of adenosinetriphosphatase-deficient islands in rat liver.
Cancer Res. 40, 4261-4264 (1980).

**16.) Kunz. W., Appel, K.E., Rickart, R, Schwarz, M.,
Stöckle, G.:**
Enhancement and inhibition of carcinogenic effectiveness
of nitrosamines. In: Primary liver tumours: H. Remmer,
H.M. Bolt, P. Bannasch and H. Popper (Eds.),
MTP Press, Lancaster, UK, 261-283 (1978).

**17.) Kunz, H.W., Tennekes, H.A., Port, R.E., Schwarz, M.,
Lorke, D, Schaude, G.:**
Quantitative aspects of chemical carcinogenesis and tumor
promotion in liver.
Environm. Health Perspect. 50, 113-122 (1983).

**18.) Kunz, H.W., Schwarz, M, Tennekes, H., Port, R.,
Appel, K.E .:**
Mechanism and dose-time response characteristics of
carcinogenic and tumour promoting xenobiotics in liver.
In: Tumorpromotoren, Erkennung, Wirkungsmechanismen und
Bedeutung. BGA-Schriften 6: Appel, K.E. und Hildebrandt,
A.G. (Eds.)
MMV Medizin Verlag München, 76-94 (1985).

19.) Kunz, H.W., Buchmann, A., Schwarz, M., Schmitt, R., Kuhlmann, W.D., Wolf, C.R., Oesch, F.:
Expression and inducibility of drug metabolizing enzymes in preneoplastic and neoplastic lesions of rat liver during nitrosamine-induced hepatocarcinogenesis.
Arch. Toxicol. 60, 198-203 (1987).

20.) Moolgavkar, S.H., Knudson, A.G.:
Mutation and cancer: A model of human carcinogenesis.
J.Natl.Cancer Inst. 66, 1037-1052 (1981).

21.) Moolgavkar, S.H.:
Model of human carcinogenesis: action of environmental agents.
Environm. Health Perspect. 50, 285-291 (1983).

22.) Nims, R.W., Devor, D.E., Henneman, J.R., Lubet, R.A.:
Induction of alkoxyresorufin O-dealkylases, epoxide hydrolase, and liver weight gain: correlation with tumor-promoting potential in a series of barbiturates.
Carcinogenesis, 8, 67-71 (1987).

23.) Pitot, H.C., Barsness, L., Goldsworthy, T., Kitagawa, T.:
Biochemical characterisation of stages of hepato-carcinogenesis after single dose of diethylnitrosamine.
Nature 271, 456-458 (1978).

24.) Pitot, H.C., Sirica, A.E.:
The stages of initiation and promotion in hepatocarcinogenesis.
Biochim. Biophys. Acta 605, 191-215 (1980).

25.) Preussmann, R., Habs, M., Habs, H., Schmähl, D.:
Carcinogenicity of N-nitrosodiethanolamine in rats at five different dose levels.
Cancer Res. 42, 5167-5171 (1982).

26.) Rabes, H.M., Szymkowiak, R.:
Cell kinetics of hepatocytes during the preneoplastic period of diethylnitrosamine-induced liver carcinogenesis.
Cancer Res. 39, 1298-1304 (1979).

27.) Rabes, H.M., Bücher, T., Hartmann, A., Linke, I., Dünnwald, M.:
Clonal growth of carcinogen-induced enzyme deficient preneoplastic cell population in mouse liver.
Cancer Res.: 42: 3220-3227 (1982).

28.) Rabes, H.M., Kerber, R., Wilhelm, R.:
Development and growth of early preneoplastic lesions induced in the liver by chemical carcinogens.
J. Cancer Res. Clin. Oncol. 106, 85-92 (1983).

29.) Rajewsky, M.F., Dauber, W., Frankenberg, H.:
Liver carcinogenesis by diethylnitrosamine in the rat.
Science 152, 82-85 (1966).

114

30.) Rotstein, J., Macdonald, P.D.M., Rabes, H.M., Farber, E.:
Cell cycle kinetics of rat hepatocytes in early putative preneoplastic lesions in hepatocarcinogenesis.
Cancer Res. 44, 2913-2917 (1984).

31.) Scherer, E., Emmelot, P.:
Kinetics of induction and growth of precancerous liver-cell foci, and liver tumour formation by diethylnitrosamine in the rat.
Eur. J. Cancer 11, 689-696 (1975).

32.) Scherer E.:
Relationship among histochemically distinguishable early lesions in multistep-multistage hepatocarcinogenesis.
Arch. Toxicol. Suppl. 10, 81-94 (1987).

33.) Schulte-Hermann, R.:
Tumor promotion in the liver.
Arch.Toxicol. 57, 147-158 (1985).

34.) Schwarz, M, Pearson, D., Port, R., Kunz, W.:
Promoting effect of 4-dimethylaminoazobenzene on enzyme altered foci induced in rat liver by N-nitroso-diethanolamine.
Carcinogenesis 5, 725-730 (1984).

35.) Schwarz, M., Buchmann, A., Pearson, D., Peres, G., Schael, S., Kuhlmann, W.D. and Kunz, H.W.:
Alterations in gene expression in preneoplastic and neoplastic hepatic lesions. In: Primary changes and control factors in carcinogenesis: Friedberg, T. (Ed.)
Dr. Braun Verlag, Wiesbaden (1986).

36.) Schwarz, M., Pearson, D., Buchmann, A., Kunz, W.:
The use of enzyme-altered foci for risk assessment of hepatocarcinogens. In: Biologically based methods for cancer risk assessment: C.C.Travis (Ed.)
NATO ASI Series, Plenum Press New York, London (1989).

37.) Solt, D.B., Farber, E.:
New priciple for the analysis of chemical carcinogenesis.
Nature (London), 263, 701-703 (1976).

38.) Weibel, E.R., Stäubli, W., Gnägi, H.R., Hess, F.A.:
Correlated morphometric and biochemical studies on liver cells; I. Morphometric model, stereological methods, and normal morphometric data for rat liver.
J.Cell Biol. 42, 68-91, 1969.

39.) Weiler, E.:
Die Änderung der serologischen Spezifität von Leberzellen der Ratte während der Carcinogenese durch p-Dimethylamino-azobenzol.
Z. Naturforschung 11b, 31-38 (1956).

40.) Weinberg, W.C., Berkwits, L., Iannaccone, P.M.:
The clonal nature of carcinogen-induced altered foci of g-glutamyl transpeptidase expression in rat liver.
Carcinogenesis, 8, 565-570 (1987)

CELL GROWTH DYNAMICS AND DNA ALTERATIONS IN CARCINOGENESIS

Samuel M. Cohen and Leon B. Ellwein

ABSTRACT

A biological model of carcinogenesis has been developed based on a two event process. The frequency of each event is dependent on the number of target cells, the frequency of their cell division, and the probability of a critical genomic error occurring with each cell division. A chemical can increase the likelihood of cancer by affecting either the rate of genomic errors (genotoxicity) or the rate of cell division, or both. Using these parameters a more rationale approach to quantitative risk assessment is proposed. Three chemicals are presented as examples of how genotoxic effects (2-acetylaminofluorene in the mouse liver, the "megamouse," ED_{01} study) and proliferative effects (sodium saccharin) influence carcinogenesis and how genotoxic and proliferation effects can interact (2-acetyla-minofluorene in the mouse bladder and N-[4-(5-nitro-2-furyl)-2-thiazolyl]formamide in the rat bladder). The influence of pharmacokinetics, metabolism, physiology, and cell kinetics on the carcinogenicity of specific compounds is illustrated, demonstrating the importance of mechanistic information in assessing potential risk.

Carcinogenesis is a complex process involving the transformation of a normal cell to a malignant cell, which then evolves into a destructive neoplasm. This process has been studied extensively in animal models to understand better the mechanisms involved. Chemicals are considered to be a major etiologic factor in the genesis of human tumors, and elimination of potential carcinogens from our environment is a major goal of our society and government. Obviously, it is best not to wait for the appearance of an increased incidence of a tumor in humans before eliminating the compound. Attempts to identify carcinogens before human exposure occurs is a major objective of animal bioassays, mechanistic studies, and short-term screens. Part of this identification process is to quantitatively estimate potential risk in humans exposed to relatively low doses of chemicals based on studies in experimental animals involving relatively high doses of the chemicals. To date, the methods utilized for these extrapolations are strictly mathematical [26]. It is increasingly becoming apparent that a more realistic approach to risk assessment, although more complicated, is to base it on a biological model of the carcinogenic process taking into account both qualitative and quantitative aspects [12,19]. Mathematical expressions in this setting are derived from the biological model.

The Biological Model

Approximately ten years ago, we developed a biological model of carcinogenesis based on a two-stage process [12]. We utilized a simulation approach as the mathematical representation of the process. Our original model was developed and based on urinary bladder carcinogenesis studies in rats; we have recently demonstrated that it is appropriate for studies in other species (the mouse) and other organs (the liver). The basics of the model are illustrated in Figure 1. Like any other model, we have made several assumptions about the carcinogenic process; these are based entirely on experimental studies on the biology of carcinogenesis. The first assumption is that the process

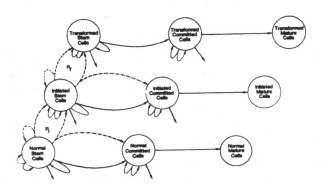

Figure 1. Possible cellular transitions. Arrows pointing down and to the right represent loss due to cell death. See text for description of model and assumptions that were made. (From Greenfield, Ellwein, and Cohen [12].)

of going from normal to transformed cells and involves two
irreversible, genetic events. By transformed, we mean
malignant cells, not benign lesions. We label the first
step initiation and the second step transformation. The
intermediate cell population is composed of initiated cells,
and includes premalignant lesions.

The second assumption is that the carcinogenic process
applies to only a select population of cells in a given
organ. We refer to these as stem cells, but in reality,
they represent multi-potential precursor cells which have
the capability of dividing and reproducing themselves. In
layered epithelia, such as the bladder or skin, this
represents a proportion of cells in the basal layer. In
solid organs, such as the liver, it is more difficult to
ascertain specifically what this population is. Clearly,
hepatocellular tumors in the liver only arise from hepatocytes or hepatocyte precursors.

A third assumption is that both genetic events can occur
only when cells are in active division, not at rest. It is
unclear at what stage of the cell cycle the critical events
occur, but they do not occur when the cell is in G_0.
Theoretically, a cell at rest could be affected, but the
genetic alteration is not fixed until a cell division
occurs. The possibility of genetic transition from a normal
to an initiated cellular state or from an initiated to a
transformed state is represented probabilistically, i.e.,
every time a cell undergoes cell division, there is a
certain probability that one of the two critical events
necessary for carcinogenesis will occur. This means that
even under control conditions, there is a certain
"spontaneous" rate of genetic error. These probabilities
are represented as P_I and P_T in the model; P_I represents the
probability of a normal cell becoming initiated during a
cell cycle, and P_T represents the probability of an
initiated cell becoming transformed during a cell cycle.

Obviously, any agent that affects the carcinogenic
process has to affect one of the variables implicit in this

model. Clearly, with classical, genotoxic chemicals, one would assume that there is an increased chance of changes in the genetic components of the cell; P_I and/or P_T would be affected. However, it is also implicit in the model that the number of cells in the normal and initiated populations, and their rate of division, will greatly affect the number of opportunities for genetic events to occur. Thus, not only could an agent increase the likelihood of cancer by causing genetic alterations, but also by increasing cell number and/or cell proliferation rates. In this presentation, we will describe three examples of how these variables affect the dose response of classical genotoxic carcinogens and also of non-genotoxic carcinogens [5,8,9,12]. Implications with respect to risk assessment for these different types of chemicals are discussed.

The biological model of Figure 1 can be viewed mathematically as a discrete-time Markov process with time-varying parameters. Unfortunately, the time-varying nature of the underlying biological process essentially precludes development of closed-form mathematical representations, and a statistical "fit" of the model to available data for purposes of parameter estimation is not feasible. On the other hand, the multiplicity and size of the cell populations makes an approach based entirely on the simulation of individual cells extravagant in terms of computer storage and running time. We utilize mathematical representations of the "mean" behavior of the cellular system to improve efficiency.

The discrete time perspective provides for representation that is eminently suited for computer modeling and analysis. In our modeling, we partition the time horizon of the study into segments on each of which the parameters are held constant; these can be as numerous as necessary to adequately reflect parameter changes brought on by complex experimental protocols. We have also found that a large number of short time segments are essential for representing the dramatic changes in proliferation rates that occur <u>in utero</u>, during the neonatal time period, and following weaning for tissues such as the bladder. Adequate representation of cellular dynamics is particularly important for chemicals where the carcinogenic influence is driven by cell proliferation rather than genotoxicity.

It is essential in the modeling process to have as much data as possible on cell population sizes and their proliferation characteristics over several time points. Also, it is essential to have tumor prevalence information at more than one time point to estimate differential effects on P_I and P_T. Estimates of the growth of tumors early in their formation is necessary since experimental prevalence data pertains to tumors of detectable size.

N-[4-(5-nitro-2-furyl)-2-thiazolyl]formamide (FANFT)

Our initial modeling efforts were based on the simulation of numerous experiments involving the urinary bladder carcinogen N-[4-(5-nitro-2-furyl)-2-thiazolyl] formamide (FANFT) in F344 rats [12,14]. FANFT is a typical carcinogen [5] since it is metabolically activated to a reactive electrophile, which forms DNA adducts, and is strongly mutagenic in short-term systems. It induces urinary bladder carcinomas when administered to rats, mice, hamsters, and dogs. It is metabolically activated by deformylation to 2-amino-4-(5-nitro-2-furyl)thiazole (ANFT) which is excreted in the urine in high concentrations. An experimental model was developed in the F344 rat that delineated the early changes and their reversibility in response to feeding FANFT at a dose of 0.2% of the diet for variable lengths of time followed by control diet; animals were also sacrificed at 26, 52, 78, and 104 weeks to determine tumor prevalences (Figure 2) [3,6,16].

Cell number estimates were based on bladder size and histopathologic evaluations. At this fixed dose, there is a prominent hyperplasia early in the process which continues for the duration of the time FANFT is administered. Mitotic rates were derived from labeling index experiments utilizing one hour pulse treatments of ^{3}H-thymidine. P_I and P_T were

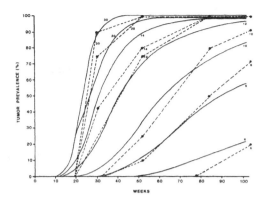

Figure 2. Comparison of results of FANFT carcinogenesis experiments in rats with those predicted by the computer model. The data points from each rat experiment are connected by a dashed line (-----) and show the proportion of rats with visible tumors at selected points in time as a result of a specified period of FANFT administration followed by control diet. The data points correspond to the times at which rats were killed in a given experiment. The numbers next to the lines represent the weeks that that group of rats was fed FANFT. The results predicted by the computer model are shown with a solid line (———). (From Greenfield, Ellwein, and Cohen [12].)

inferred by comparing model predictions with experimental tumor prevalence data at the different sacrifice time points. It was found that FANFT markedly altered the initiation probability but did not affect the transformation probability. This is consistent with subsequent experiments where rapidly proliferating, initiated bladder epithelial cells appear to be unresponsive to FANFT administration [7].

Our modeling efforts then addressed the issue of dose response in the more traditional sense [5,14]. Different doses of FANFT were administered in the diet, and tumor prevalences were determined after one and two years (Table 1).

TABLE 1. Effect of Dose on the Urinary Bladder
Carcinogenicity and Proliferative Response to FANFT
in Male F344 Rats

Dose of FANFT (%)	Bladder Carcinoma Prevalence (%)		Labeling Index (% ± S.D.)
	1 yr.	2 yr.	After 10 Weeks
0.2	--	100[a]	2.02±0.84[a]
0.1	56[a]	100[a]	2.21±0.74[a]
0.05	29[a]	87[a]	0.84±0.36[a]
0.01	0	0	0.11±0.05
0.005	0	0	0.08±0.06
0.001	0	0	0.08±0.06
0.0005	0	0	---
0	0	0	0.10±0.07

a. Significantly different from control group (0% FANFT)
at $p < 0.05$ by the Fisher exact or Student's t-test. Data
are from Hasegawa et al. [14].

There is nearly a 100% prevalence of tumors at the
higher doses of FANFT, but there is a sharp drop in tumor
probability at a dose of 0.01% of the diet and below. Since
there is no observed effect at the lower doses, one might
conclude that a no-effect level exists. However, by
modeling it is demonstrated that tumors are produced, but
at a frequency that is not detectable given the number of
animals utilized. It has been well demonstrated that
administration of FANFT at these lower doses results in
FANFT being excreted in the urine, and that there is a
co-carcinogenic effect of FANFT at these low doses [21].
What then is happening that might explain this marked change
in slope of the tumor prevalence curve at the lower doses
compared to the higher doses?

As indicated above, not only does FANFT interact with
DNA, but it also increases cell number (hyperplasia) and
mitotic activity in the urothelium. We have determined the
dose response for the cell proliferation effects of FANFT,
as reflected in the labeling indices shown in Table 1. The
increased proliferation rates and hyperplasia are seen only

at the higher doses at which tumors result, whereas at the lower doses, there is no increased proliferation. Thus, the increased probability of bladder tumors in the animals fed FANFT at the higher doses is due to a synergistic combination of the genotoxic effect of FANFT (increased P_I) and an increase in the cell number and proliferation rate in the epithelium. Presumably there is a proportional effect on P_I, at the lower doses, but without the increased proliferative stimulus, the prevalence of tumors falls below

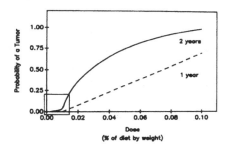

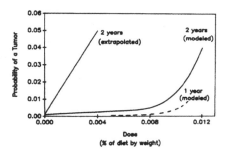

Figure 3. Modeled dose-response for rats fed FANFT for 30 weeks and examined after one year (-----) or two years (————). The lower Figure is an expansion of the lower left corner of the upper Figure. The solid line labeled 2 years (extrapolated) represents a straight line extrapolation on the basis of observed tumor prevalence at FANFT doses of .05, .10 and .20%

the level of experimental detectability. On the basis of modeling analyses, we demonstrated quantitatively this dependence on target organ proliferative effects as the biological explanation for the two slopes in the dose-response curve. This is illustrated in Figure 3. Although FANFT is clearly a genotoxic carcinogen, proliferative effects markedly alter the tumor dose-response and considerably affect the extrapolation to low doses.

2-Acetylaminofluorene (AAF)

AAF is another classical carcinogen which has been evaluated extensively regarding its metabolism and mechanism of action in carcinogenesis [1]. In addition, an enormous database is available on the dose response effects of AAF, particularly at low doses. This database comes from the so-called "mega-mouse" experiment performed at the National Center for Toxicological Research and published in 1980 [24]. This experiment involved more than 24,000 female mice administered AAF at several doses, and was designed to detect a prevalence of tumors greater than 1% over background (ED_{01}). Unlike FANFT, which is primarily a urinary bladder carcinogen, AAF has two major target sites in female mice, the liver and the urinary bladder. Interestingly, the dose response for AAF was quite different in the two tissues. In the liver, the tumor response was apparently linear with respect to dose, including down to very low doses (30 ppm). In contrast, no increased prevalence of bladder tumors was seen at the lower doses, but an increased prevalence was seen at the higher doses, suggesting the possibility of a threshold effect.

This experiment provides a database which is extremely useful for our modeling purposes. In addition to the fact that several doses were used, there was a scheduled sacrifice at several time points during the experiment. In some groups the administration of AAF was discontinued, and the mice were killed at subsequent times. Considerable data are also available on the hyperplastic effects in the bladder; little data is available on cellular proliferation within the liver.

Information has also become recently available with regard to the metabolism of AAF in mice and DNA adduct levels in target tissues. Beland et al. have demonstrated that the dose response for DNA adduct formation in the liver and the bladder is approximately linear, extending below doses utilized in the ED_{01} study [1,2]. Also, Beland and others have demonstrated that the metabolic activation of AAF in the liver results in high levels of DNA adducts in normal hepatocytes, but metabolism does not occur (or is considerably less) in hyperplastic foci [1,15,18]. It is likely, therefore, that AAF affects the initiation probability, P_I, in the liver, but not the transformation probability, P_T. In contrast, it would be anticipated that P_I and P_T are affected in the bladder since AAF is activated in the liver and transported as the N-glucuronide to the urine where it is cleaved to the reactive metabolite [1,17].

These metabolic and adduct studies provide the ultimate in pharmacokinetic modeling since target dose relationships and DNA adduct formation are compared directly to administered dose. In the liver there is no evidence for cellular proliferation at the doses of AAF utilized in the ED_{01} study, whereas considerable hyperplasia is produced in the bladder at the higher AAF doses of this experiment. The hyperplastic response in the bladder varies with respect to time for each given dose [24].

To model the effects of AAF on the liver, we made the assumption that P_I increases linearly with respect to dose based on the observed linear relationship between DNA adduct level and dose. Since there are no proliferative effects and no anticipated effect on P_T, it is not surprising that tumor response as a function of administered dose is approximately linear at these low doses of AAF. A major factor in modeling of the liver is the fact that there is a high prevalence of liver tumors in the control mice, those not administered AAF. This is very helpful in inferring spontaneous levels of P_I and P_T for the liver. Utilizing background cell numbers and proliferation rates, as well as baseline values for P_I and P_T, the control dose response curve is readily obtainable.

In the bladder, modeling results are considerably different. For one, the spontaneous incidence of bladder tumors is nearly zero (inferred baseline values for P_I and P_T are 3×10^{-5}.) It remains at this low level until a dose of 75 ppm of the diet, when it begins to rise sharply [24]. Again, similar to the findings with FANFT, the apparent non-response at the low doses is not truly a threshold since the expected prevalence is below the 1% detection level of the experimental design. Also, it is unlikely that the low-dose response is at control levels since DNA adducts are formed at the lower doses as demonstrated by Beland et al [13,15]. In that study, as discussed above, the DNA adduct levels in the bladder were linear with dose as in the liver. As with FANFT, detectable tumor prevalence coincides with those doses for which there is an increased level of AAF-induced cell proliferation. Without the proliferative effect at the lower doses, the genotoxic effect on the bladder, on P_I and P_T, is inadequate to generate a detectable tumor prevalence. However, when this genotoxic effect is combined with increased cell proliferation at the higher doses, a large detectable increase in tumor prevalence occurs. We have calculated that the expected bladder tumor prevalence would have been 6% at 33 months had there only been a genotoxic effect at 150 ppm. Similarly, we have calculated that the expected tumor response at 150 ppm would be 14% if there were only a cell proliferation effect of AAF on the bladder. Together, tumor prevalence is 100% due to the synergism between cell proliferation and genotoxic effects.

The above analyses deal with ED_{01} results utilizing the continuous feeding of different AAF doses. In addition, several groups were administered AAF for 9, 12, or 15 months and then examined at 18 and 24 months for tumor prevalence. These interval experiments are extremely helpful in distinguishing between alternative mechanistic possibilities. For example, in the liver, these discontinuous studies provide considerable support for an effect only on P_I without any effect on P_T. In contrast, they also provide considerable support that in the bladder there is an effect on P_I and P_T rather than only on P_I.

Again, a classical carcinogen can be seen to have genotoxic effects, but the dose response with respect to tumor prevalence is greatly altered by the proliferative effects of the chemical. In the case of AAF in mice, there is no proliferative effect at the doses used in the ED_{01} study in the liver, whereas there is a considerable proliferative effect in the bladder. This produces two very different tumor dose response curves. Extrapolations to lower doses are also affected by such influences.

Sodium Saccharin

Sodium saccharin is a considerably different type of chemical than FANFT or AAF. It does not fit the paradigm of a classical carcinogen. It is not metabolized to an active electrophile (it is actually nucleophilic), it does not react with DNA, and it is not mutagenic in short term bioassays except in a few instances at extremely high doses [5,8,10]. When it is administered at high doses in the diet to rats beginning after weaning, a significant increase in bladder tumors is not produced. However, if it is administered in the diet beginning at conception through gestation and lactation and then after weaning for the remainder of the rat's life, a significantly increased frequency of bladder tumors results, particularly in males [10]. It has subsequently been demonstrated that the critical time begins at the time of parturition, and that there is no additional influence with exposure during gestation [22]. Thus, the period from parturition to weaning is an extremely critical time period. Also, in these studies there is a very steep dose response curve. Extremely high doses are administered in these studies. Most of the two-generation studies have utilized 5% of the diet as the administered dose, and a positive response has been found at doses as low as 3% of the diet. No increased effect has been seen at the level of 1% of the diet [10,22].

In addition to carcinogenicity in two-generation studies at these high doses, sodium saccharin has been shown to promote bladder tumor carcinogenesis [4,5,10,19] following initiation by FANFT, N-butyl-N-(4-hydroxybutyl)nitrosamine (BBN),N-methyl-N-nitrosourea (MNU), or freeze ulceration. Also, sodium saccharin is demonstrated to be co-carcinogenic when administered as 5% of the diet with 0.005% FANFT [21]. If it does not have a genotoxic effect, how does sodium saccharin induce bladder cancer?

Following sodium saccharin administration to rats at high doses in the diet [5,8,10,11], there is increased cell proliferation of the urothelium. This occurs within one week of administering sodium saccharin and continues for the period of time that it is administered in the diet. There is also a dose response, with increased proliferation at a level of 2.5% and above, but not at lower dietary doses [20]. Is this increased cell proliferation adequate to explain the various complex protocols in which sodium saccharin results in bladder tumor induction in the rat? We have extensively modeled all of the above referenced experiments and have found that the increased cell proliferation at the different time points readily explains the carcinogenic effects of sodium saccharin in these protocols [8]. The complexities in such protocols with respect to cell proliferation rates are illustrated in Figure 4.

Inherent in these studies is the implication that doses of sodium saccharin which result in increased cell proliferation have the potential of increasing tumor probability. What is the effect at lower doses? The apparent answer to this question is that at doses of 1% and below there is no increased cell proliferation, and therefore, an increased incidence of bladder cancer is not expected [10]. However,

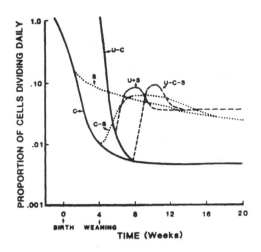

Figure 4. Stem cell mitotic rates (reflecting normal and initiated cells combined) approximated from series of labeling index measurements. The different experimental regimens are identified by C = control diet; S = sodium saccharin; U = ulceration. (From Ellwein and Cohen [8].)

to base this apparent threshold effect on a firmer foundation, we must understand the mechanism of action of sodium saccharin in the rat and how this mechanism might apply to the human situation.

We have extensively been examining the mechanism of action of sodium saccharin in the male rat, the only susceptible species and the most susceptible gender. We have found that administering saccharin as the calcium salt or as the acid form does not produce the same response as sodium saccharin when administered at similar levels in the diet [13]. Initially, we hypothesized that this was due to differences in solubility, and therefore differences in excretion levels in the urine. However, similar concentrations of saccharin are excreted in the urine following the administration of any salt form. Although the saccharin levels are similar following administration of the various salt forms, there are differemces in pH, sodium concentration, and the concentration of numerous other factors in the urine. We have demonstrated that the chemical form of saccharin is not altered by any of these ionic alterations in the urine [27], and we have also demonstrated that there is no cell receptor for saccharin [10]. These differences in the urinary and proliferative effects between the sodium salt and the acid have been observed for ascorbate and other organic acids [5,10].

More recently, we have observed that, following the administration of sodium saccharin at high levels of the diet to male rats, there is formation of a large amount of a finely precipitating material and formation of crystals which have an appearance somewhat different from those seen in control rats when observed by scanning electron micro-scopy [4]. This precipitated material and the different crystals are composed largely of silicon-containing materials, possibly silica. A major component of urinary crystallization in animals is the protein matrix. The greater the amount of protein in the urine, the more likely the precipitate will form. It is well known that male rat urine has considerably more protein than female rat urine or urine from other species [25]. Also, this precipitate and crystallization is inhibited by decreasing pH; it occurs to a small or no extent below pH 6.5 [23]. Low pH also appears to inhibit the induction of urothelial proliferation following saccharin feeding [10,13]. In preliminary experiments, we demonstrated that this proteinaceous and silicon-containing precipitate forms to a very slight extent in female rats, and there is also few of the silicon-containing crystals in female rats. In male and female mice, none of this precipitate occurs, and extremely few silicon-containing crystals are formed in the urine.

This correlates with the absence of tumors in these species. It would appear that the mechanism of sodium saccharin on the urothelium in the male rat is dependent on the formation of the silicon-containing precipitate and/or crystals, and this is dependent on high levels of protein in the urine and a neutral or higher pH. It would, there-fore, be unlikely for this material to appear at lower concentrations of sodium saccharin or at lower concen-trations of protein in the urine. The implication is that this is a threshold effect in the male rat, and that the threshold can be determined chemically by determining the level of saccharin in the urine at a given urinary pH and protein to form these precipitates. Extrapolating this to the female rat suggests a considerably reduced risk of formation of this precipitate, because of the lower protein in the urine. In mice, this precipitate does not appear at doses of sodium saccharin in the diet as high as 7.5%. Since this precipitate and crystal formation would not be expected to form in humans at the ingested dose levels of saccharin in humans, no effect would be expected to occur in humans. Essentially, our hypothesis is that high doses of sodium saccharin and a certain urinary milieu are required for a proliferative response to occur in the urothelium. Only if this proliferative response occurs is there a chance for developing bladder cancer in these animals. A threshold effect is hypothesized and a rational basis for extrapolation to humans is thus available.

With these non-genotoxic compounds, it is important in extrapolating from high dose animal experiments to low doses in humans to take into account the mechanism of action. Mathematical extrapolation that does not account for biological mechanism would end up with a falsely high level of estimated risk. Risk assessment in this instance is dependent on demonstrating that a biological basis can be derived for the action of a non-genotoxic compound, and that there is a threshold level for its effect.

Classification of Chemicals

Based on these models and others which are available in the literature with different tissues and species, we have classified chemicals somewhat differently from current practice. Essentially, we divide chemicals into two broad categories: those which are genotoxic and those which are non-genotoxic. Genotoxic compounds can only be modeled with respect to a dose response relationship if both the genotoxic and cell proliferation effects are known. Clearly, the dose at which cell proliferation effects become influential is important, since this greatly alters the slope of the dose response curve.

In contrast, non-genotoxic compounds by definition do not have a direct genotoxic effect: the P_I and P_T remain at background levels. These chemicals act by increasing cell proliferation of either normal cells and/or initiated cells, which leads to an increased likelihood of the initiation and/or transformation event occurring, even though P_I and P_T remain at background, control levels. These compounds are further divided into two separate categories: those which act via a cell receptor, either membrane or intracytoplasmic, and those which do not act via a cell receptor, but generally act by means of either direct mitogenicity or cytotoxicity and consequent regenerative hyperplasia. These compounds will have their effect only if they increase cell proliferation of normal and/or initiated cells. Depending on their relative effects on normal and initiated cells, different dose and time response curves would be anticipated.

We do not categorize chemicals using terms such as initiators, promoters, complete carcinogens, or a number of other terms which are currently used to classify compounds. We believe that there is a better way of approaching the classification of potential chemical carcinogens, and thereby providing a more rational basis for the extrapolation of risk to humans based on animal experiments. Clearly, the limited results of a typical 2-year bioassay in animals are inadequate for proper modeling and extrapolation.

ACKNOWLEDGEMENTS
    This work was supported in part by USPHS Grants CA32513, CA44886, CA28015, and CA36727 from the National Cancer Institute and by grants from the State of Nebraska Department of Health and from the International Life Sciences Institute - Nutrition Foundation. We gratefully acknowledge the assistance of Jan Leemkuil and Ginni Philbrick in the preparation of this manuscript.

REFERENCES

1. Beland, F.A. Metabolic activation of aromatic amine carcinogens in vitro and in vivo. J. Univ. Occupat. Environ. Hlth., in press.

2. Beland, F.A., Fullerton, N.F., Kinouchi, T., and Poirier, M.C. DNA adduct formation during continuous feeding of 2-acetylaminofluorene at multiple concentrations. In: Methods for Detecting DNA Damaging Agents in Humans: Applications in Cancer Epidemiology and Prevention, H. Bartsch, K. Hemminki and I.K. O'Neill, Eds., IARC, Lyon, pp. 175-180, 1988.

3. Cohen, S.M., Arai, M., Jacobs, J.B., and Friedell, G.H. Promoting effect of saccharin and DL-tryptophan in urinary bladder carcinogenesis. Cancer Res. 39: 1207-1217, 1979.

4. Cohen, S.M., Cano, M., Garland, E.M., and Earl, R.A. Silicate crystals in the urine and bladder epithelium of male rats fed sodium saccharin. Proc. Am. Assoc. Cancer Res. 30: 205, 1989.

5. Cohen, S.M., and Ellwein, L.B. Cell growth dynamics in long-term bladder carcinogenesis. Toxicol. Lett. 43: 151-173, 1988.

6. Cohen, S.M., Jacobs, J.B., Arai, M., Johansson, S., and Friedell, G.H. Early lesions in experimental bladder cancer: Experimental design and light microscopic findings. Cancer Res. 36: 2508-2511, 1976.

7. Cohen, S.M., Murasaki, G., Fukushima, S., and Greenfield, R.E. Effect of regenerative hyperplasia on the urinary bladder:Carcinogenicity of sodium saccharin and N-[4-(5-nitro-2-furyl)-2-thiazolyl] formamide. Cancer Res. 42: 65-71, 1982.

8. Ellwein, L.B., and Cohen, S.M. A cellular dynamics model of experimental bladder cancer: Analysis of the effect of sodium saccharin in the rat. Risk Analysis 8: 215-221, 1988.

9. Ellwein, L.B., and Cohen, S.M. Comparative analyses of the timing and magnitude of genotoxic and nongenotoxic cellular effects in urinary bladder carcinogenesis. In: Biologically-Based Methods for Cancer Risk Assessment, C.C. Travis, Ed., Plenum Publ. Corp, pp. 181-192, 1989.

10. Ellwein, L.B., and Cohen, S.M. The health risks of saccharin revisited. In: CRC Reviews in Toxicology, in press.

11. Fukushima, S., and Cohen, S.M. Saccharin-induced hyperplasia of the rat urinary bladder. Cancer Res. 40: 734-736, 1980.

12. Greenfield, R.E., Ellwein, L.B., and Cohen, S.M. A general probabilistic model of carcinogenesis: Analysis of experimental urinary bladder cancer. Carcinogenesis 5: 437-445, 1984.

13. Hasegawa, R., and Cohen, S.M. The effect of different salts of saccharin on the rat urinary bladder. Cancer Lett. 30: 261-268, 1986.

14. Hasegawa, R., Cohen, S.M., St.John, M., Cano, M., and Ellwein, L.B. Effect of dose on the induction of urothelial proliferation by N-[4-(5-nitro-2-furyl)-2-thiazolyl]formamide and its relationship to bladder carcinogenesis in the rat. Carcinogenesis 7: 633-636, 1986.

15. Huitfeldt, H.S., Hunt, J.M., Pitot, H.C., and Poirier, M.C. Lack of acetylaminofluorene-DNA adduct formation in enzyme-altered foci of rat liver. Carcinogenesis 9: 647-652, 1988.

16. Jacobs, J.B., Arai, M., Cohen, S.M., and Friedell, G.H. A long-term study of reversible and progressive urinary bladder cancer lesions in rats fed N-[4-(5-nitro-2-furyl)-2-thiazolyl]formamide. Cancer Res. 37:2817-2821, 1977.

17. Kadlubar, F.F., Miller, J.A., and Miller, E.C. Hepatic microsomal N-glucuronidation and nucleic acid binding of N-hydroxy arylamines in relation to urinary bladder carcinogenesis. Cancer Res. 37: 805-814, 1977.

18. Lai, C.-C., Miller, J.A., Miller, E.C., and Liem, A. N-Sulfooxy-2-aminofluorene is the major ultimate electrophilic and carcinogenic metabolite of N-hydroxy-2-acetylaminofluorene in the livers of infant male C57BL/6J x C3H/HeJ F1 (B6C3F1)mice. Carcinogenesis 6:1037-1045, 1985.

19. Moolgavkar, S.H. Biologically motivated two-stage model for cancer risk assessment. Toxicol. Lett. 43:139-150, 1988.

20. Murasaki, G., and Cohen, S.M. Effect of dose of sodium saccharin on the induction of rat urinary bladder proliferation. Cancer Res. 41: 942-944, 1981.

21. Murasaki, G., and Cohen, S.M. Co-carcinogenicity of sodium saccharin and N-[4-(5-nitro-2--furyl)-2-thiazolyl]formamide for the urinary bladder. Carcinogenesis 4: 97-99, 1983.

22. Schoenig, G.P., Goldenthal, E.I., Geil, R.G., Frith, C.H., Richter, W.R., and Carlborg, F.W. Evaluation of the dose response and in utero exposure to saccharin in the rat. Fd. Chem. Toxicol. 23: 475-490, 1985.

23. Schreier, C.J., and Emerick, R.J. Diet calcium carbonate, phosphorus and acidifying and alkalizing salts as factors influencing silica urolithiasis in rats fed tetraethylorthosilicate. J. Nutr. 116: 823-830, 1986.

24. Staffa, J.A., and Mehlman, M.A. (Eds.) Innovations in cancer risk assessment (ED_{01} Study). J. Environ. Path. Toxicol. 3: 1-246, 1980.

25. Swenberg, J.A., Short, B., Borghoff, S., Strasser, J., and Charbonneau, M. The comparative pathobiology of $_{2u}$-globulin nephropathy. Toxicol. Appl. Pharmacol. 97: 35-46, 1989.

26. VanRyzin, J. The assessment of low-dose carcinogenicity: Discussion. Biometrics 38(Supplement): 130-139, 1982.

27. Williamson, D.S., Nagel, D.L., Markin, R.S., and Cohen, S.M. Effect of pH and ions on the electronic structure of saccharin. Fd. Chem. Toxicol. 25: 211-218, 1987.

Two Mutation Model for Carcinogenesis: Relative Roles of
Somatic Mutations and Cell Proliferation in Determining
Risk

Suresh H. Moolgavkar, Georg Luebeck and Mathisca de Gunst

ABSTRACT: Two experimental data sets are analyzed within
the framework of a two-event model for carcinogenesis.   In
the first, the number and size distribution of altered
hepatic foci, which are thought to be premaligant lesions,
are analyzed as functions of dose of an administered agent
(N-Nitrosomorpholine, NNM).   Definitions of initiation and
promotion potencies are proposed.   Results of the analysis
indicate that NNM is a strong initiator and a weak
promoter.   In the second, the time to appearance and the
probability of malignant lung tumors in rats exposed to
radon are analyzed as functions of total exposure and rate
of exposure.   The results indicate that fractionation of
exposure increases the lifetime probability of tumor, and
that the efficiency of fractionation can be explained by
the relative effects of radon daughters on the mutation
rates and the kinetics of growth of initiated cells.

1.   INTRODUCTION

The two-event model for carcinogenesis developed by
Moolgavkar and Venzon [8] and Moolgavkar and Knudson [9]
brings together multistage models of carcinogenesis and
Knudson's ideas on recessive oncogenesis.   The model has
been shown to be consistent with epidemiologic and
experimental data.   Recently, there has been some interest
in the use of this model for quantitative cancer risk
assessment [e.g., 14, 18].   Attractive features of the
model are that it takes explicit account of cell kinetics
and that the concepts of initiation and promotion are
clearly definable within its framework.   The purpose of
this paper is to illustrate the usefulness of the model for
the analysis of experimental carcinogenesis data.   Analyses
of two data sets are presented.   These analyses show rather
nicely that both cell kinetics and somatic mutations are
important in carcinogenesis.

Before discussing the examples, we briefly review the
model and its implications for initiation and promotion.

2. THE MODEL

The model views malignant transformations as resulting from two rare (rate-limiting), irreversible and heritable cellular events that lead to abrogation of growth control. As in Knudson's [7] model for embryonal tumors, these events may be thought of as mutations at homologous sites (antioncogene or tumor suppressor gene loci) on the genome. However, the model does not require such interpretation. See Knudson (this volume) for possible biological interpretations of the two events.

For a mathematical development of the model we assume that once-hit, intermediate or initiated, cells are generated from normal susceptible cells in the tissue of interest according to a non-homogeneous Poisson process with intensity $\nu(s)X(s)$, where $\nu(s)$ is the first "mutation" rate at time (age) s, and $X(s)$ is the number of normal susceptible cells. It is convenient to think of $\nu(s)$ as a locus-specific mutation rate; however that assumption is not necessary. Indeed, Clifton (this volume) presents evidence that the initiating event, at least in radiation carcinogenesis, is relatively common, perhaps too frequent to be a locus-specific mutation.

Once an intermediate or initiated cell is created from a normal cell, it may divide into two intermediate cells, or die (or differentiate), or divide into one intermediate cell and one malignant cell (cell that has sustained the second critical mutation). The instantaneous rates for these events will be denoted by $\alpha(s)$, $\beta(s)$ and $\mu(s)$, respectively. The quantities of interest to be derived from the model are the hazard (incidence) and probability functions for the appearance of a malignant cell and expressions for the number and size of premalignant foci. Papillomas in mouse skin-painting experiments, enzyme-altered foci in hepatocarcinogenesis experiments and adenomatous polyps in the human colon are examples of premalignant foci.

The mathematical details are presented elsewhere and will not be repeated here [see 2, 10].

Action of environmental agents

Within the framework of the model, agents may alter the risk of cancer by affecting either the probability of generating intermediate cells from normal cells or the probability of generating malignant cells from intermediate cells, or both. Each of these probabilities clearly depends upon the corresponding "mutation" rate (ν or μ) and on the kinetics of tissue proliferation.

Within the framework of the model, an initiator is any agent that increases ν, the rate of the first critical event. If the nature of the two critical events is the same, then an agent that increases ν will also increase μ and act as a "complete" carcinogen. Note that this does not mean that the agent has promoting activity. A promoter is any agent that increases the rate of intermediate cell division or decreases the rate of intermediate cell death (or differentiation). Such an agent usually increases ($\alpha-\beta$) resulting in an increase in the population of intermediate cells, which are the targets for the second event. In addition, if a promoter increases α, then it must also increase μ, if the second event is a mutation (because increased cell division implies increased mutation rates per unit time). The implications of this model of initiation-promotion have been discussed in detail in previous publications [e.g., 9]. Some facts that should be kept in mind for the analysis of enzyme altered foci to be discussed next are: (1) In any population of cells undergoing a "birth-death" process, the probability of extinction is non-zero. Thus, initiated cells have a positive probability of dying out before they give rise to intermediate foci. In fact, the (asymptotic) probability that an initiated cell (together with all its daughters) will become extinct is β/α. In our analysis of NNM generated intermediate foci (see below) we estimated β/α to

be as high as 0.9. (2) Suppose that there is a certain
threshold (say n_o cells) below which foci are not visible.
Then any focus ultimately becomes visible or extinct.
(3) The number and size of visible intermediate foci depend
upon both the rate of initiation and the parameters α and β
determining the growth kinetics of initiated cells. In
particular, the number of foci depends upon the rate of
initiation and on the probability of extinction of
initiated cells. Thus, it is impossible to infer from the
number of foci alone the mode of action of an agent (i.e.,
whether an agent is an initiator or a promoter). (4) It is
easy to show, under fairly general conditions, that the
distribution of sizes of intermediate foci depends only
upon the promoting activity of an agent and not on its
initiating activity. Put in another way, if an agent
affects ν without affecting α or β, then exposure to the
agent will not change the size distribution of intermediate
foci.

3.   ANALYSIS OF ADENOSINE TRIPHOSPHATASE (ATPase)
      DEFICIENT LIVER FOCI IN RATS EXPOSED TO N-
      NITROSOMORPHOLINE (NNM)

      One of the attractive features of the rodent liver
model for chemical carcinogenesis is the appearance of
phenotypically altered clonal populations of cells which
are characterized by changes in the expression of a variety
of biochemical markers, and at least some of which are
believed to be precursor lesions for malignant tumors [3,
5, 6, Pitot et al. this volume]. Although considerable
effort has been expended at quantifying the number and size
distribution of these enzyme-altered foci, no systematic
analyses of the data so obtained have been undertaken. In
this paper we outline a method of analysis based on the
two-event model. The first and second events represent
"initiation" and "progression", respectively, and clonal
expansion of intermediate (initiated) cells represents
"promotion".

The data we analyze in this example were kindly provided to us by Dr. Michael Schwarz. Rats were exposed to varying concentrations (0, 0.1, 1, 5, 10, 20, 40 ppm) of NNM in their drinking water. The following information was available on each animal: the concentration of NNM in the drinking water, the age of the animal in days when it was sacrificed, the number of ATPase deficient foci observed in a two-dimensional section of the liver and the area of the section, the radii in microns of each of the observed foci. The objective of the analysis was to study the effect of NNM on the parameters of the model.

The relevant formulae for the number and the size distribution of altered foci can be found in a recent paper by Dewanji et al. [2]. Two expressions from that paper are of particular importance. The number of non-extinct altered foci at age t has a Poisson distribution with mean $\Lambda(t)$ given by

$$\Lambda(t) = \int_0^t \nu(s)X(s)[1-p(t;s)]ds \qquad (1)$$

where $p(t;s)$ is the probability that an altered cell generated at time s, together with all its daughters, is extinct by time t. Let $W(t)$ represent the number of cells in a randomly chosen (non-extinct) focus at time t. Then, for any m=1,2,...,

$$\text{Prob}[W(t)=m] = [\Lambda(t)]^{-1}\int_0^t \nu(s)X(s)[1-p(t;s)] \times$$
$$[1-(\alpha/\beta)p(t;s)][(\alpha/\beta)p(t;s)]^{m-1} ds \qquad . \qquad (2)$$

Note that $p(t;s)$ depends upon α and β, but not upon ν.

If information on the number and size distribution (in terms of the number of cells in each focus) were directly available for each rodent liver, then expressions (1) and (2) could be used to form a likelihood for the data and the parameters of interest estimated by maximizing this likelihood. However, the data are not available in this form. Instead, for each rodent liver, we know the number of transections (of foci) above a certain critical size observed under the microscope in a given cross-sectional area of the liver, and the radius in microns of each of these transections. Thus, three problems need to be addressed before expressions (1) and (2) can be used: 1) The observed transections are in two dimensions, whereas the expressions derived from the model are in three. Thus, we need a method to bridge the two. 2) Only transections above a certain critical size are observable. 3) Information on size of foci in microns must be translated into information on size in terms of number of cells, or vice versa.

Traditionally, the first problem has been addressed by using standard stereological procedures to compute the number and size-distribution of three-dimensional objects from two-dimensional observations. Generally, methods proposed by Fullman or Saltykov are used [1, 13]. Unfortunately, the Fullman and Saltykov procedures have some unpleasant properties. The variance of the Fullman estimator is infinite [16]. Saltykov's method requires binning of sizes and often leads to negative estimates in some of the bins, especially when the number of observed transections is small. Thus, we chose to translate our model-derived three-dimensional quantities into two-dimensional quantities using a formula due to Wicksell [17]. In the current analysis, X represents the number of cells per cubic centimeter, and the parameters m, α, β are assumed to be independent of time (but dependent on dose of NNM). Technical details regarding likelihood construction and maximization can be found in a recent paper [12].

Results and discussion

The objective of the analysis is to obtain estimates of mX, which measures the rate of initiation, and of (α-β), which measures the net growth rate of initiated (ATPase deficient) cells, as functions of dose of NNM. In Figure 1 we present these estimates together with the regression

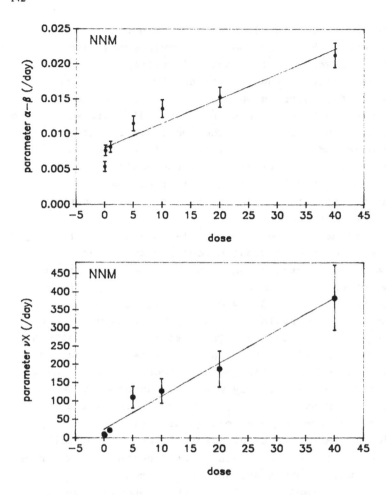

Figure 1. Parameter estimates (and 95% confidence
intervals) plotted against dose of NNM in parts
per million. The parameter νX estimates the
number of altered cells generated per day per
cubic centimeter of liver. The parameter (α-ß)
is the net rate of growth of altered cells.

lines through them. Initiation and promotion potencies can
now be defined as the quotient of the slope and the
intercept of the appropriate regression line. Thus,
potency measures the proportionate increase of νX or $(\alpha-\beta)$
over background per unit of dose. These potencies depend
upon the units in which dose is measured; however, once the
potencies are computed in a given system of units, it is an
easy matter to express them in any other system of units.

For the data analyzed here the equations of the
regression lines are

$$\nu X(d) = 23.08 + 9.04d$$

and $(\alpha-\beta)(d) = 0.008022 + 0.0003522d$

Thus, the initiation potency is $9.04 \div 23.08 = 0.392$ per
ppm and the promotion potency is $0.0003522 \div 0.008022 =$
0.044 per ppm, and the analysis indicates that NNM is a
strong initiator and a weak promoter.

The estimates of νX range from approximately 10 (in
the 0 dose group) to approximately 380 (in the 40 ppm dose
group) per day per cubic centimeter of liver. These
numbers translate, respectively, to 3,650 and 140,600
initiated cells per year per cubic centimeter. In
comparison, the model predicts approximately 100 and 1,700
nonextinct foci per cubic centimeter in dose groups 0 and
40, respectively. Thus, the vast majority of initiated
cells do not generate foci. This makes intuitive sense.
In any tissue in growth equilibrium, cell division rates
must be approximated by cell death rates. Thus, any
initiated cell must have almost as great a probability of
dying as of dividing. The net growth rate, $(\alpha-\beta)$, of
altered cells ranges from approximately 0.0054 per cell per
day in the control group to approximately 0.021 per cell
per day in the 40 ppm dose group. These values correspond
to doubling times of approximately 128 and 33 days,
respectively.

See the papers by Schwarz et al. and Pitot et al. in this volume for discussion of related topics.

4.   RADON AND LUNG CANCER IN RATS

We are indebted to Dr. Fred Cross, Battelle Pacific Northwest Laboratories, for making these data available. Detailed analyses of the data will appear elsewhere [11].

With respect to the carcinogenic potential of high LET radiation, some of the most important questions revolve around the issue of dose-response. In particular, there is evidence that fractionation of high LET radiation leads to an increased lifetime probability of malignant tumors [4, 15]. Here we present a summary of an analysis of experimental data on radon-induced lung tumors in rats within the framework of the two mutation model. We find that fractionation of exposure results in increased lifetime probability of malignant tumor, and that a biological explanation of this observation can be given in terms of the relative effect of radon on the mutation rates and the kinetics of intermediate cells.

In the experiment, 1,797 rats were subjected to several different lifetime exposures to radon. Within each total exposure group, there were several different exposure-rate regimens. During the course of the experiment, any animal that died or was moribound was necropsied. In some subgroups serial sacrifices were conducted. A total of 326 rats developed histologically verified malignant lung tumors. In the opinion of the pathologist, tumors did not cause death of the animals and hence were treated as being "incidental" in the statistical analysis.

The following information was available on each animal: the exact age (in days) when exposure to radon daughters was begun, the exposure rate in working levels months per week (WLM/w), the age at which exposure was stopped, age at death or sacrifice, and presence or absence of malignant lung tumor. Exposures on all animals started sometime during the 14th week of life, and all animals were followed until sacrifice or death.

Let P(t) be the probability of tumor by age t for some particular exposure-rate regimen. Then, the survivor function S(t) = 1-P(t) and the hazard function h(t) = -S'(t)/S(t). Since the tumors are incidental, the contribution to the likelihood by an animal that died (or was sacrificed) at age t is P(t) if it had a tumor or S(t) if it was free of tumor. The full likelihood is the product of these terms over all animals.

The objectives of the analysis are to estimate the first and second mutation rates, $\nu(d)$ and $\mu(d)$, respectively, and the rates of intermediate cell proliferation, $\alpha(d)$, and differentiation (or death), $\beta(d)$, as functions of the exposure rate, d, of radon. We estimated these parameters by maximizing the likelihood. Note that the parameters are functions of the rate of radon exposure while radon is being administered. Before and after exposure the parameters revert to background levels.

The parameter α is thought to be approximately 10 per cell per week (C.L. Saunders, personal communication; precisely, cell division rates have been measured in adenomas which are probably intermediate lesions in the rat lung). Thus, as functions of the exposure rate, d, the parameters are explicitly modelled as follows

$$\alpha(d) = 10.00$$
$$\alpha-\beta(d) = a + b \; ln(1+d)$$
$$\nu(d) = c_1 \; \exp(p_1 \; ln(1+d))$$
$$\mu(d) = c_2 \; \exp(p_2 \; ln(1+d))$$

Statistical results can be found in the paper by Moolgavkar et al. [11].

Table 1.   Observed and expected number of tumors generated by model. Exposure refers to total exposure of radon daughters received by the animals. Each group of animals was exposed to a particular exposure rate regimen d. Thus, for example, the exposure rate regimen in group 6 ranges from 0.53 to 0.63 $Jhm^{-3}w^{-1}$ (150 to 180 working level months per week). For each group we show, within each exposure rate and total exposure category, from top to bottom, the number of animals in the experiment, the number of malignant tumors observed, and the expected number of malignant tumors generated by the model.

		0	1.1	2.2	4.5	8.8	1.8	35 Jhm^{-3}	TOTAL
Group expos:		0	320	640	1280	2500	5000	10000WLM	
	0.018$Jhm^{-3}w^{-1}$								
1	d 5 WLM/w	32	128	0	0	0	0	0	160
	Obs.	0	16	0	0	0	0	0	16
	Expec.	0.83	18.35	0	0	0	0	0	19.18
	0.18$Jhm^{-3}w^{-1}$								
2	d 50 WLM/w	96	127	64	32	32	32	0	383
	Obs.	1	8	17	17	22	25	0	90
	Expec.	2.29	14.89	14.08	13.91	19.68	22.11	0	89.96
	1.8$Jhm^{-3}w^{-1}$								
3	d 500 WLM/w	81	169	105	74	72	71	76	648
	Obs.	0	12	5	11	20	27	43	118
	Expec.	1.34	7.14	7.24	9.03	18.95	31.04	47.37	112.12
	0.91$Jhm^{-3}w^{-1}$								
4	d 260 WLM/w	64	0	0	0	0	128	0	192
	Obs.	1	0	0	0	0	68	0	69
	Expec.	1.47	0	0	0	0	65.59	0	67.05

		0	0.07	0.14	0.28	0.56	1.1	2.2Jhm^{-3}	TOTAL
Group expos:		0	20	40	80	160	320	640 WLM	
	0.18$Jhm^{-3}w^{-1}$								
5	d 50 WLM/w	18	18	18	18	18	18	18	126
	Obs.	0	0	0	0	0	0	0	0
	Expec.	0.08	0.10	0.12	0.16	0.23	0.46	0.97	2.12
	0.53-0.63$Jhm^{-3}w^{-1}$								
6	d 150-180 WLM/w	0	0	0	0	0	96	0	96
	Obs.	0	0	0	0	0	8	0	8
	Expec.	0	0	0	0	0	6.55	0	6.55
	0.18-0.60$Jhm^{-3}w^{-1}$								
7	d 50-170 WLM/w	32	0	0	0	0	0	160	192
	Obs.	0	0	0	0	0	0	25	25
	Expec.	0.52	0	0	0	0	0	19.98	20.50

Results and discussion

The results of the analysis are presented in Tables 1 and 2. Table 1 presents the number of malignant tumors observed in each of the exposure rate categories, together with the expected number generated by the model. This latter number was calculated as follows. For each animal we used the estimated parameters to compute the model generated lifetime probability, based on time of death or sacrifice, that the animal had a tumor. These probabilities were then summed over all the animals in a particular exposure category to yield the expected number of animals with tumor in that category.

We note that the mathematical form of the model does not allow us to identify the number of normal cells separately from the first mutation rate.

As can be seen from Table 1, the two-event model provides a good description of radon-induced lung tumors in rats. Because the parameters of the model are functions of the exposure rate at which radon is administered, it is possible to model the appearance of tumors continuously throughout the lifespan of an animal, before, during and after radon exposure. This, of course, is more desirable than monitoring the appearance of tumors at one particular age.

The analysis of these data indicate that radon strongly increases the first mutation rate and also increases the net rate of intermediate cell growth. Radon has a smaller effect on the second mutation rate, suggesting that the nature of the two mutational events is different.

As shown in Figure 2, the lifetime probability of tumor is higher with fractionation of a given total exposure. A biological explanation of this phenomenon within the context of this model is as follows. During exposure to radon, there is an increase in the pool of intermediate cells due to direct mutation of the normal cells. These cells in turn grow into clones the sizes of

148

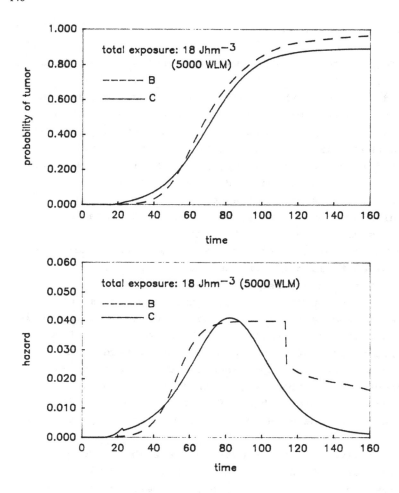

Figure 2. Probability of tumor (upper panel) and hazard
(incidence) of tumor (lower panel) associated
with a total exposure to radon daughters of 18
Jhm^{-3} (5000 WLM), at two exposure-rate regimens,
plotted against age in weeks. B: 0.18 $Jhm^{-3}w^{-1}$
(50 WLM/w); C: 1.8 $Jhm^{-3}w^{-1}$ (500 WLM/w). Curves
were generated using the parameter estimates in
Table 2. Exposure to radon daughters was assumed
to start at 14 weeks of age. Note that, after
exposure ceases, the hazard functions ultimately
decline with time.

which depend on continued exposure to radon. This increase in the number of intermediate cells of course increases the probability that one of these cells will suffer the second mutation and become malignant. For a given total exposure, a low exposure rate over a prolonged period is more efficient at producing malignant transformation because the mutation rates and the intermediate cell proliferation rate are sublinear functions of exposure rate.

Finally, we note here that, from the statistical point of view, time-dependent exposure patterns are very naturally incorporated into an analysis using the model described here. With the traditional statistical models this is not so easy.

Table 2.   Results of maximum likelihood fitting of model to the radon data. Total number of animals = 1,797; total number of tumor bearing animals = 326. Standard errors computed from the diagonal entries of the inverse of the observed information matrix.

ln likelihood = -538.3

Parameter	m.l.e[a]	s.e.[b]	Wald 95% C.I.[c] m.l.e.+1.96 s.e.		Likelihood-based 95% C.I.	
a	7.259×10^{-2}	2.358×10^{-2}	$(2.638 \times 10^{-2},$	$1.188 \times 10^{-1})$	$(3.412 \times 10^{-2},$	$1.312 \times 10^{-1})$
b	2.018×10^{-2}	4.629×10^{-3}	$(1.110 \times 10^{-2},$	$2.925 \times 10^{-2})$	$(1.278 \times 10^{-2},$	$3.253 \times 10^{-2})$
c_1	1.015×10^{-1}	4.436×10^{-2}	$(1.452 \times 10^{-2},$	$1.884 \times 10^{-1})$	$(4.716 \times 10^{-2},$	$3.200 \times 10^{-1})$
c_2	3.379×10^{-6}	1.710×10^{-6}	$(2.706 \times 10^{-8},$	$6.730 \times 10^{-6})$	$(5.998 \times 10^{-7},$	$7.338 \times 10^{-6})$
p_1	0.8113	5.127×10^{-2}	(0.7108,	0.9118)	(0.7133,	0.9124)
p_2	0.1150	1.571×10^{-1}	(<0,	0.4229)	(<0,	0.4572)

[a] maximum likelihood estimate
[b] standard error
[c] confidence interval

5. CONCLUDING REMARKS

A simple model for carcinogenesis with two rate-limiting steps ("intiation" and "progression") separated by clonal expansion ("promotion") of "initiated" cells is consistent with the experimental and epidemiologic data. In this paper we have discussed two examples which illustrate dramatically the interplay between somatic mutations and cell kinetics in determining cancer risk. For other examples, see the papers by Cohen and Ellwein, Chen and Moini, and Grosser and Whittemore in this volume.

A purpose of the model is to focus attention on the type of data that are needed for a rational approach to risk assessment. It should come as no surprise that quantitative information on fundamental biological parameters, such as stem cell number, division, death and mutation rates, is required. This information is clearly going to be difficult to obtain, and even when the information does become available, it remains to be shown that the model has predictive, rather than just descriptive, power.

6. ACKNOWLEDGEMENTS

Supported by grant CA-47658 from NIH and contract DOE DE-FG06-88GR60657. Dr. de Gunst was supported by a NATO fellowship from the Netherlands Organization for Scientific Research.

REFERENCES

1. Campbell HA, Zu, YD, Hanigan MH, Pitot HC:
    Application of quantitative stereology to the
    evaluation of phenotypically heterogeneous enzyme-
    altered foci in the rat liver. JNCI 76:751-767,
    1986.

2. Dewanji A, Venzon DJ, Moolgavkar SH: A stochastic
    two-stage model for cancer risk assessment. II.
    The number and size of premalignant clones. Risk
    Analysis 9:179-187, 1989.

3. Emmelot P, Scherer E: The first relevant cell
    stage in rat liver carcinogenesis: A quantitative
    approach. Biochimica et Biophysica Acta 605:247-
    304, 1980.

4. Fry RJM: Experimental radiation carcinogenesis:
    What have we learned? Radiation Research 87:224-
    239, 1981.

5. Goldfarb S, Pugh TD: Enzyme histochemical
    phenotypes in primary hepatocellular carcionmas.
    Cancer Research 41:2092-2095, 1981.

6. Goldsworthy TL, Henigan MH, Pitot HC: Models of
    hepatocarcinogenesis in the rat -- contrasts and
    comparisons. CRC Critical Review of Toxicology
    17:61-89, 1986.

7. Knudson AG: Mutation and cancer: Statistical
    study of Retinoblasoma. Proceedings of the
    National Academy of Sciences, USA 68:820-823, 1971.

8. Moolgavkar SH, Venzon DJ: Two-event models for
    carcinogenesis: Incidence curves for childhood and
    adult tumors. Math Biosci 47:55-77, 1979.

9. Moolgavkar SH, Knudson AG Jr: Mutation and cancer:
    A model for human carcinogenesis. JNCI 66:1037-
    1052, 1981.

10. Moolgavkar SH, Dewanji A, Venzon DJ: A stochastic
    two-stage model for cancer risk assessment. I.
    The hazard function and the probability of tumor.
    Risk Analysis 8:383-392, 1988.

152

11. Moolgavkar SH, Cross FT, Luebeck EG, Dagle RE: A two-mutation model for radon-induced lung tumors in rats. Radiation Research, in press.

12. Moolgavkar SH, Luebeck EG, de Gunst M, Port RE, Schwarz M: A method for the quantitative analysis of enzyme-altered foci in rat hepatocarcinogenesis experiments I: Single agent regimen. Submitted.

13. Nychka D, Pugh TD, King JH, Koen H, Wahba G, Chover J, Golfarb S: Optimal use of sampled tissue sections for estimating the number of hepatocellular foci. Cancer Research 44:178-183, 1984.

14. Thorslund TW, Brown CC, Charnley G: Biologically motivated cancer risk models. Risk Analysis 7:109-119, 1987.

15. Ullrich RL: Tumor induction in BALB/c mice after fractionated or protracted exposures to fission-spectrum neutrons. Radiation Research 97:587-597, 1984.

16. Watson GS: Estimating functionals of particle size distributions. Biometrika 58:483-490, 1971.

17. Wicksell DS: The Corpuscle problem, Part I. Biometrika 17:87-97, 1925.

18. Wilson JD: Risk assessment reappraisals. Science 240:1126, 1988.

Cancer Dose-Response Models Incorporating Clonal Expansion

Chao W. Chen and Assad Moini

Abstract

Under the assumption that a malignant tumor develops through a sequence of steps (normal cells → initiated cells/foci → nodules → tumors) two classes of mathematical models of carcinogenesis that have a potential to be used for cancer dose-response modeling are discussed. The two classes of models considered are (1) a general version of the two-stage model by Moolgavkar and colleagues [12, 13], henceforth called the MVK model, and (2) a clone process model derived from Tucker [21]. These two classes of models incorporate essentially the same biological information but in different ways and offer a conceptual contrast between the two differing approaches. The objectives of this paper are to (1) highlight issues and problems that arise in using biologically based dose-response models to predict cancer risk and (2) discuss how parameters in the models could be estimated using auxiliary information.

We have also demonstrated that use of an approximate form of the MVK model may lead to a biologically unrealistic implication of the model and an underestimation of risk at low doses when parameters are estimated from bioassay data.

1. Introduction

Most scientists who are involved in risk assessments would agree that it is desirable to incorporate available information on biological mechanisms into quantitative risk assessment; the problem, however, is to determine how the available mechanistic information may be incorporated. The search for mechanisms of carcinogenesis has been extremely diverse. For instance, some mechanistic studies have focused on events that might be involved in the initiation of transformation such as specific changes in the cell membrane or cytoskeleton or in the expression of specific growth factors, enzymes, or oncogene products. While these studies are fundamental to the advancement of cancer research, this type of information alone is not sufficient for dose-response modeling, because it does not account for the dynamic nature of cellular progression from the normal to malignant stage. To model the tumor growth and, eventually, incorporate this into the dose-response relationship, one currently must be content with the mathematical description of the carcinogenic process at the cellular level rather than at the organelle or molecular level. For instance, Moolgavkar and colleagues [12, 13] have postulated that tumorigenesis proceeds through two stages (i.e., from normal to initiated cells and from initiated to malignant cells) via clonal growth of initiated cells (I-cells). If clonal growth is part of the tumorigenic process, then the rate of clonal growth may be a rate-limiting step for tumor formation. There is evidence that an agent with only initiating capability would not induce a significant tumor increase unless the I-cells can be proliferated by the agent itself or by the host condition. For instance, in their studies of vinyl chloride (VC) on female Wistar rat liver, Laib et al. [9] found that continuous exposure of adult rats to VC did not result in an increase in either the area of foci or the incidence of liver tumors over that of controls, even though the number of small ATPase-deficient foci were increased when compared with control animals. On the other hand, animals that were exposed to VC in the early lifetime, when the liver cell proliferation rate is high, resulted in an increase in both the area of preneoplastic foci and in the incidence of hepatocellular carcinomas. These observations suggest that a biologically based dose-response (BBDR) model should consider not only number of I-cells but also both the size (i.e., number of cells) and frequency of foci/nodules. For another discussion of this theme, see also the paper by Moolgavkar et al. in this volume.

In this paper, two mathematical dose-response models that incorporate information on cellular dynamics (growth kinetics and progression from one step to another) are applied to two sets of data obtained from animal studies on VC via inhalation and benzo[a]pyrene (B[a]P) by skin painting. The two models considered are a general version of the two-stage model by Moolgavkar and colleagues [12, 13] and a model derived from Tucker [21]. These two models use essentially the same biological information but in different ways. The conceptual contrast of these models offers an opportunity to demonstrate how the same biological information may be incorporated in different ways to construct a BBDR model. The use of two different sets of data in this paper serves to illustrate the kind of biological information that is useful for BBDR modeling.

2. Mathematical Description of Models

Model 1: A class of MVK models.

Given below are some mathematical derivations that are necessary for our discussion. For further investigations about the generalized MVK model, one should consult papers by Serio [16], Tan and Gastardo [18], Tan and Brown [19], and Moolgavkar et al. [14].

Basic assumptions:

1. The number of normal cells (N-cells) is deterministic in time, t, following some nonexponential growth function, N(t). Although it is mathematically more general to assume a birth-death process for N-cells which results in exponential growth, we find that the use of deterministic nonexponential growth is biologically more realistic.

2. The number of I-cells produced by N-cells is a Poisson process with a production rate, $\mu_1(t)N(t)$, where $\mu_1(t)$ is the rate of initiation for an N-cell at time, t.

3. An I-cell follows a nonhomogeneous birth-death process with a mutation. That is, an I-cell can divide into two I-cells, die, or divide into an I-cell and a malignant cell. Denote the birth, death, and mutation rates at time, t, by $b(t)$, $d(t)$, and $\mu_2(t)$, respectively.

4. Cells go through the processes independent of each other.

Let $Y(t)$ and $Z(t)$ be random variables representing the number of I-cells and malignant cells, respectively, at time, t. Let $\phi(y, z \mid t_0, t)$ be the probability generating function (p.g.f.) of $[Y(t), Z(t)]$ given that the process starts with a single I-cell and zero malignant cells at time $t = t_0$ (i.e., $Y(t_0) = 1$, $Z(t_0) = 0$). It is straightforward to derive Kolmogorov's forward equation for ϕ:

$$\frac{\partial \phi}{\partial t} = \{y^2 b(t) + y(z - 1)\mu_2(t) + d(t) - y[b(t) + d(t)]\} \frac{\partial \phi}{\partial y} \tag{1}$$

with $\phi(y, z \mid t_0, t_0) = y$.

Incorporating the Poisson process of producing I-cells from N-cells as described in assumption 2, let $\psi(y, z \mid t_0, t)$ be the p.g.f. for $[Y(t), Z(t)]$, given $N(t_0)$, $Y(t_0) = 0$ and $Z(t_0) = 0$. It can be shown [14] that ψ is given by

$$\psi(y, z \mid t_0, t) = \exp\left\{ \int_{t_0}^{t} \mu_1(s)N(s)[\phi(s, t) - 1]ds \right\} \tag{2}$$

Thus, the survival function of time to the first malignant cell since time, t_0, as defined in Moolgavkar et al. [14] is $\psi(1, 0 \mid t_0, t)$, which has the hazard function given by

$$h(t_0, t) = -\frac{d}{dt} \log \psi(1, 0; t_0, t).$$

156

By equating the time to the first malignant cell as time to tumor, $h(t_0, t)$ is interpreted as the hazard function of the time to tumor.

Using the fact that $[\frac{\partial}{\partial t} \phi(y, z \mid t_0, t)] \mid_{y=1,z=0} = \frac{\partial}{\partial t} [\phi(y, z \mid t_0, t) \mid_{y=1,z=0}]$, Eq. 1 yields

$$\frac{\partial \phi(1, 0; s, t)}{\partial t} = -\mu_2(t) \frac{\partial \phi(y, z; s, t)}{\partial y} \mid_{y=1,z=0} \tag{3}$$

Thus,

$$h(t_0, t) = -\int_{t_0}^{t} \mu_1(s)N(s) \frac{\partial \phi(1, 0; s, t)}{\partial t} ds \tag{4}$$

$$= \mu_2(t) \int_{t_0}^{t} \mu_1(s)N(s) \left[\frac{\partial \phi(y, z; s, t)}{\partial y} \mid_{y=1,z=0} \right] ds$$

To calculate hazard function $h(t_0, t)$, the solution for ϕ in Eq. 1 must be obtained. The analytic solution for ϕ is difficult to obtain except for some special cases. Two special cases are given here.

Case 1: Homogeneous processes, $b(t) = b$, $d(t) = d$, $b > d$, and $\mu_2(t) = \mu_2$.

Under these assumptions, $\phi(y, z; t_0, t) = \phi(y, z; t - t_0)$ is given by (see e.g., Serio [16])

$$\phi(y, z; t - t_0) = \frac{y_1 - y_2 \left(\dfrac{y - y_1}{y - y_2} \right) \exp[b(y_1 - y_2)(t - t_0)]}{1 - \left(\dfrac{y - y_1}{y - y_2} \right) \exp[b(y_1 - y_2)(t - t_0)]} \tag{5}$$

where $y_1 < y_2$ are two real roots of the quadratic equation

$$bx^2 - [b + d + \mu_2(1 - z)]x + d = 0.$$

Thus, the exact hazard function of time to tumor is given by

$$h(t_0, t) = \mu_2 \int_{t_0}^{t} \mu_1(s)N(s)S(t, s)ds \tag{6}$$

where

$$S(t, s) = \frac{\partial}{\partial y} \phi(1, 0 \mid t - s)$$

$$= \frac{(y_2 - y_1)^2 \exp[b(y_2 - y_1)(t - s)]}{\{(1 - y_1) + (y_2 - 1)\exp[b(y_2 - y_1)(t - s)]\}^2} \tag{7}$$

where $y_1 < y_2$ are two real roots of $bx^2 - (b + d + \mu_2)x + d = 0$.

Approximate Hazard Function:

When $\mu_2 \to 0$,

$$\tilde{h}(t_0, t) = \mu_2 \int_{t_0}^{t} \mu_1(s)N(s)\exp[(b - d)(t - s)]ds; \, b > d. \tag{8}$$

Eq. 8 was used by Thorslund et al. [20] to construct a BBDR model. It has been pointed out by Moolgavkar et al. [14] that the use of such an approximated formula is not appropriate when the tumor incidence is high. As shown below, the use of the approximate formula (Eq. 8) may not only give a wrong risk estimate but may also be biologically misleading. To demonstrate why the approximate formula may result in biologically misleading implications, we consider a case where an exposure to a carcinogen affects the initiation rate, μ_1.

Some Special Results:

Let $h_0(t)$ and $h_1(t)$ be hazard rates for control and carcinogen-exposed groups, respectively. Define an excess hazard rate to be $h_e(t) = h_1(t) - h_0(t)$, where $h_e(t) \geq 0$ for every t. Some special results which are used to demonstrate that the approximate hazard function, $\tilde{h}(t)$, can be misleading are given below.

Result 1: Assume that the $N(t) = N$, $\mu_1(t) = \mu_1$, and an exposure to a carcinogen has an effect of increasing μ_1 to $\mu_1(1 + \delta)$ with $\delta > 0$. If the exposure occurs in time interval $[x_0, x_1]$, then the excess hazard rates at $t > x_1$ is given by

$$h_e(t \mid x_0, x_1) = \delta\mu_1\mu_2N \frac{(y_2 - y_1)}{b(y_2 - 1)} [A(x_1) - A(x_0)] \tag{9}$$

158

where

$$A(x_i) = \frac{1}{(1 - y_1) + (y_2 - 1)\exp[b(y_2 - y_1)(t - x_i)]} \; ; i = 0, 1.$$

Derivation of Eq. 9 proceeds as follows:

From Eq. 6, for $t > x_1$,

$$h_1(t \mid x_0, x_1) = \mu_1\mu_2 N[\int_0^{x_0} S(t, \tau)d\tau + (1 + \delta)\int_{x_0}^{x_1} S(t, \tau)d\tau + \int_{x_1}^{t} S(t, \tau)d\tau]$$

where S is given by Eq. 7. Therefore, the excess hazard for $t > x_1$ is

$$h_e(t \mid x_0, x_1) = \delta\mu_1\mu_2 N \int_{x_0}^{x_1} S(t, \tau)d\tau.$$

Eq. 9 follows by noting that the integration of Eq. 7 has the form $A(x_i)$ multiplying by a factor $(y_2 - y_1)/[b(y_2 - 1)]$.

Result 2: Under the same conditions given in Result 1, the excess hazard after an acute exposure at age x_0 is given by

$$h_e(t \mid x_0) = \frac{\delta\mu_1\mu_2 N(y_2 - y_1)^2\exp[b(y_2 - y_1)(t - x_0)]}{\{(1 - y_1) + (y_2 - 1)\exp[b(y_2 - y_1)(t - x_0)]\}^2} \; ; t > x_0 \qquad (10)$$

Derivation of Eq. 10 proceeds as follows:

By letting $x_1 = x_0 + \varepsilon$ in Eq. 9, we note that

$$h_e(t \mid x_0) = \lim_{\varepsilon \to 0} \frac{1}{\varepsilon} h_e(t \mid x_0, x_0 + \varepsilon)$$

$$= \frac{\delta\mu_1\mu_2 N(y_2 - y_1)}{b(y_2 - y_1)} \frac{d}{dx_0} [-A(x_0)].$$

Under the same conditions given in Result 1, it is trivial to write down equations corresponding to Eq. 9 and Eq. 10 when the approximate hazard rate function (Eq. 8) is used. For instance, for an acute exposure, the approximate excess hazard rate is given by

$$\tilde{h}_e(t \mid x_0) = \delta\mu_1\mu_2 N exp[(b - d)(t - x_0)]; \; t > x_0 \tag{11}$$

In both Results 1 and 2, the excess hazard decreases to zero over time after termination of exposure, while the approximate hazard continues to increase exponentially. The fact that the approximate hazard function, $\tilde{h}_e(t)$, is always exponentially increasing, irrespective of biological implication of the data, is very serious when it is used to interpret data from a study or to estimate parameters of the model from data. This behavior is contradictory to the well-known observation in epidemiological studies that risk of death due to tumor, as a function of age, can be increasing or decreasing. It is also inconsistent with the inherent property of the MVK model which can increase and decrease, depending of the choice of parameters in the model. Using Eq. 10 and Eq. 11, Figure 1 demonstrates a case when an acute exposure is involved.

The above demonstration also suggests that if the growth rate, $g = b - d$, is dose-dependent, the parameter μ_2 could be drastically underestimated when the approximate formula is used to estimate parameters on the basis of bioassay data, because the hazard rate is dominated by the exponential term which is a function of the growth rate, g (see Example 1 for estimation of μ_2).

Case 2: Piece-wise homogeneous processes.

Divide [0,t] into k subintervals, $J_i = [t_{i-1}, t_i)$, $i = 1, 2, 3..., k - 1$ and $J_k = [t_{k-1}, t_k]$, with $t_0 = 0$ and $t_k = t$. Assume that, for s in J_i, $\mu_1(s) = \mu_{1i}$, $\mu_2(s) = \mu_{2i}$, $b(s) = b_i$, and $d(s) = d_i$.

Let $\phi_i(t) = \phi_i(y, z; t)$, $t \epsilon J_i$, be the p.g.f. of $[Y(t), Z(t)]$, given that the process starts with a single I-cell at t_{i-1}. Then $\phi_j(t)$ has the same form as Eq. 5 in Case 1, above; y_1 and y_2 are now two real roots of the quadratic equation

$$b_i x^2 - [b_i + d_i + \mu_{2i}(1 - z)]x + d_i = 0.$$

The hazard function for a malignant cell is shown by Tan and Gastardo [18] to be

$$h(t) = \sum_{j=1}^{k} [\mu_{1j}\mu_{2j} \int_{t_{j-1}}^{t_j} N(s)m_j(t_j - s)ds] \prod_{i=j+1}^{k} m_i(t_i - t_{i-1}) \tag{12}$$

160

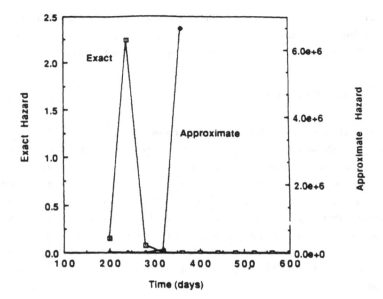

FIGURE 1. Comparison of excess hazard rates, calculated by exact (Eq. 10) and approximate (Eq. 11) formulas, for animals exposed to a single dose of carcinogen, which increases the initiation rate from μ_1 to $\mu_1(1 + \delta)$, at the age of 200 days. Parameters used for the calculation are $\delta = 1$, $\mu_1 = 3.0 \times 10^{-8}$, $b = 0.113$, $d = 0.003$, $\mu_2 = 3.0 \times 10^{-8}$, and $N = 2,800 \times 10^6$. Note that the scale for approximate hazard is much larger than the scale for exact hazard. For this reason, approximate hazards over time interval (200, 300) are hardly noticeable.

where, $\prod\limits_{i=j+1}^{k} m_i(t_i - t_{i-1}) = 1$, when $j = k$, and

$$m_j(\tau) = \exp[\int_0^t 2b_j\phi_j(1, 0; s)ds - (b_j + d_j + \mu_{2j})\tau]$$

or equivalently

$$m_j(\tau) = \frac{(y_{2j} - y_{1j})^2 \exp[b_j(y_{2j} - y_{1j})\tau]}{\{(1 - y_{1j}) + (y_{2j} - 1)\exp[b_j(y_{2j} - y_{1j})\tau]\}^2};$$

$y_{1j} < y_{2j}$ are two real roots of $b_i x - (b_i + d_i + \mu_2)x + d_i = 0$.

This model, with $k = 2$, is used to calculate the probability of cancer in the VC example. In predicting cancer risk with this model, the assumption is made, as by Moolgavkar and colleagues [12, 13], that a tumor develops instantaneously upon the occurrence of a malignant cell.

Model 2: A Clone Process Model.

A model that considers the dynamics of foci, nodules, and malignant tumors can be developed by using results from Tucker [21] and the theory of stochastic birth-death processes. For convenience, we consider a simple model that only involves foci (or nodules) and tumors. The model can be easily generalized to any number of preneoplastic entities, including both foci and nodules.

Let $Y(t)$ and $W(t)$ be the number of foci and the number of tumors, respectively, at time t. Assume that the formation of $Y(t)$ is a Poisson process with a time-dependent intensity, $\lambda(t)$. Each of the foci has a probability of $a_1(t)h + o(h)$ of changing into a tumor and a probability of death $\beta_1(t)h + o(h)$ during $(t, t + h)$. Once a tumor is formed, it will not disappear. All foci and tumors are assumed to act independently. The p g.f. of $[Y(t), W(t)]$, with $Y(0) = 0$ and $W(0) = 0$, is given by Tucker [20] to be

$$\phi(y, w; t) = E[y^{Y(t)} w^{W(t)}]$$

$$= \exp[(y - 1)\theta_1(t) + (w - 1)\theta_2(t)]$$

where $\theta_i(t)$, $i = 1, 2$ are expected values of $Y(t)$ and $W(t)$, respectively. They satisfy the following system of equations

$$\frac{d}{dt}\theta_1(t) = -[a_1(t) + \beta_1(t)]\theta_1(t) + \lambda(t) \tag{13}$$

$$\frac{d}{dt}\theta_2(t) = a_1(t)\theta_1(t). \tag{14}$$

The initial conditions are $\theta_i(0) = 0, i = 1, 2$.

Thus, the hazard rate of the tumor is given by

$$h(t) = -\phi'(1, 0; t)/\phi(1, 0; t)$$

$$= \frac{d}{dt}\theta_2(t)$$

$$= a_1(t)\theta_1(t).$$

If formation rate, $\lambda(t)$, of foci can be obtained directly from the laboratory, then the model does not require the assumption that cells act independently, an assumption that may be violated in view of the fact that cells are known to communicate among themselves.

When direct laboratory foci formation rate, $\lambda(t)$, is not available, we propose a heuristic procedure that utilizes results similar to those in Dewanji et al. [6] to estimate the rate of foci formation. It should be noted, however, that the proposed procedure requires the assumption of cell independence.

As in Model 1, it is assumed that the number of primary I-cells at time, t, is a Poisson variable with intensity, $\mu_1(t)N(t)$. Assume that an I-cell is subjected to a nonhomogeneous birth-death process with the birth and death rates given by $b(t)$ and $d(t)$, respectively. A focus is defined as a collection of I-cells descended from a primary I-cell. The probability of death at time, t, for a focus born at $t = 0$, can be shown to be given by the following equation, a well-known result in the theory of the birth-death process [5],

$$P_0(t) = 1 - \frac{1}{\exp[\gamma(t)] + \int_0^t b(\tau)\exp[\gamma(\tau)]d\tau}$$

where

$$\gamma(t) = -\int_0^t [b(s) - d(s)]ds.$$

The probability that a focus has size, k, at time, t, is given by

$$P_k(t) = [1 - P_0(t)][1 - A(t)][A(t)]^{k-1}, k \geq 1 \text{ where}$$

$$A(t) = 1 - \exp[\gamma(t)][1 - P_0(t)].$$

Therefore, the probability that a nonextinct focus has a size, k, is given by

$$N_k(t) = [1 - A(t)][A(t)]^{k-1}, \text{ a geometric distribution.}$$

Assume that a focus becomes detectable when it contains m or more cells. The probability that a nonextinct focus becomes detectable at time, t, is

$$D_m(t) = \sum_{i=m} N_i(t) = [1 - P_0(t)][A(t)]^{m-1} \tag{15}$$

and the number of detectable foci/nodules at time, t, is a Poisson variable with mean

$$F(t) = \int_0^t \mu_1(s)N(s)D_m(t - s)ds. \tag{16}$$

Therefore, the Poisson intensity function (i.e., rate of formation of detectable foci/nodules), $\lambda(t)$, is given by

$$\lambda(t) = \frac{d}{dt} F(t)$$

$$= \mu_1(t)N(t)D_m(0) + \int_0^t \mu_1(s)N(s)D_m'(t - s)ds$$

$$= \int_0^t \mu_1(s)N(s)D_m'(t - s)ds$$

where $D_m'(t)$ is the derivative of $D_m(t)$ with respect to t.

The rates, $\alpha_1(t)$ and $\beta_1(t)$, must be estimated directly from incidence data, with or without assumption of dose dependency. This model can be easily generalized to more than two variables and is applied to the skin-painting data in the second example.

In the second example, $Y(t)$ and $W(t)$ represent, respectively, the numbers of papillomas and carcinomas at time t. In this example, the value α_1 is interpreted as the mean transition rate for a papilloma (in a population of papillomas) to convert into a carcinoma. It is not appropriate to assume that all papillomas have an identical rate converting to carcinomas because size of papillomas and, perhaps, some other factors may affect the potential for their conversion into carcinomas. The death rate, β_1, for a papilloma is assumed to be zero.

3. Examples

Example 1: Vinyl Chloride

It should be noted that this example is not to be interpreted as a quantitative risk assessment for VC. The main objectives of the example are to (1) illustrate how data on foci can be used to estimate parameters in a BBDR model and (2) highlight the problems and issues that arise from the use of a BBDR model in a quantitative risk assessment.

Data Base

Animal data on VC include liver-tumor incidence induced via inhalation and ingestion exposure, pharmacokinetic and metabolic information and percent of foci area, average area per focus, and frequency of foci per cm^2 in liver. Studies in Wistar rats showed that VC induced tumors only when animals are exposed to VC in early lifetime, a period characterized by rapid liver growth. Although the data are far from ideal for constructing a BBDR model, these data offer an opportunity to demonstrate how biological information can be incorporated into a quantitative risk assessment.

Data in Tables 1 and 2 are reconstructed from graphs in Laib et al. [8, 9, 10]. Table 1 gives number of foci per cm^2 and the corresponding estimated average radii of foci observed at 121 days after birth for female Wistar rats, exposed to 2,000 ppm of VC, 8 hours per day, 5 days per week, in various lifetime periods. Table 2 gives data similar to Table 1, except that animals were sacrificed at the end of the exposure, which was continuous from day 1 after birth for different durations. In both Tables 1 and 2, the number of foci per liver is calculated from the number of foci per cm^2 by using the method in Pugh et al. (1983). The most useful auxiliary data for parameter estimation are the number and distribution of the size of foci over time. However, the distribution of the size of foci is not available and cannot be estimated. See the paper by Moolgavkar et al. in this volume for a further discussion of this issue.

Table 1. Induction of ATPase-deficient foci in liver after exposure of female Wistar rats to 2,000 ppm VC, 8 hours per day, 5 days per week[a]

Exposure period (days)	Foci per cm^2	Estimated average focal radius (cm)	Estimated foci per liver[b]
1-11	6	0.0690	278
1-17	7	0.0667	334
1-47	15	0.0242	1,992
1-83	40	0.0110	12,747

[a] Animals were sacrificed at 121 days after birth.

[b] Number of foci per cm^3 is calculated by Pugh et al.[15] as being given by $\dfrac{n_2}{\pi(r^2 - \varepsilon^2)^{0.5}}$ where n_2 is the number of foci per cm^2, r is the average radius of foci, and $\varepsilon = 0.00384$ cm is the minimum radius of a focus that can be reliably detected. The volume of rat liver is assumed to be 10 cm^3.

SOURCE: Laib et al. [8, 9, 10].

Table 2. Induction of ATPase-deficient foci for female Wistar rats exposed to 2,000 ppm VC via inhalation, 8 hours per day, 5 days per week[a]

Exposure period (days)	Foci per cm^2	Estimated average focal radius (cm)	Estimated foci per liver
1-28	15	0.0127	3,951
1-42	18	0.0143	4,154
1-56	28	0.0196	4,632
1-70	56	0.0303	5,924

[a] All animals were sacrificed at the end of exposure.

SOURCE: Laib et al. [8, 9, 10].

Parameter Estimation

1. Table 3 gives the number of hepatocytes in Wistar rats up to 95 days, taken from the Biological Data Book [1] (pp. 99-100). These data can be adequately described by a Gompertz function which has the form,

$$N(t) = N_0 \exp\{ \frac{\beta}{\alpha} [1 - \exp(-\alpha t)]\},$$ where α and β are positive constants. (17)

Table 3. Number of hepatocytes in Wistar rats

Age (days)	No. of cells in liver ($\times 10^6$)
7	228
17	687
35	1,310
95	2,655

SOURCE: Altman and Dittmer [1].

The least square estimates for the parameters are

$N_0 = 142 \times 10^6, \alpha = 3.93 \times 10^{-2}$, and $\beta = 0.118$.

Note that $N(t)$ approaches about $2,850 \times 10^6$ cells for an adult animal. $N(t)$ is used in Example 1 for number of normal cell population.

2. According to Laib et al. [9], the liver grew at a high rate in young rats. Therefore, it is assumed in our example that an I-cell is subject to a piece-wise homogeneous birth-death process; the birth and death rates are given by $b(t) = b_1$ and $d(t) = d_1$, respectively, before the age of 50 days and $b(t) = b_2$ and $d(t) = d_2$ beyond 50 days. It is assumed that a focus becomes detectable when it contains nine or more cells. The number of foci per liver in Tables 1 and 2 are used to estimate parameters $\mu_1, b_i, d_i, i = 1, 2$, using Eq. 16, with $\mu_1(t) = \mu_1$, when animals were exposed to VC, and $\mu_1(t) = \mu_0$, when animals were not exposed to VC. To reduce the number of parameters to be estimated, it is assumed that $b_1 = Lb_2$ and $d_1 = Ld_2, L > 1$. The resultant least square estimates are as follows,

$\mu_1 = 1.236 \times 10^{-7}$, $b_1 = 5.214 \times 10^{-1}$, $b_2 = 9.388 \times 10^{-5}$, $d_1 = 1.256 \times 10^{-1}$, and $d_2 = 2.262 \times 10^5$.

Note that in actual calculation, parameters estimated are $g_1 = b_1 - d_1$, $r = b_1/d_1$ and L. They uniquely determine the specific parameter values b_i and d_i for $i = 1, 2$. The value $\mu_1 = 1.236 \times 10^{-7}$ is the initiation rate when animals were exposed to 2,000 ppm VC. In the calculation, the spontaneous rate is assumed to be negligible when compared with μ_1, an initiation rate due to a high exposure concentration of VC. It is estimated that there are about two detectable foci per animal in the control group at 121 days. Using this data point and the above birth-death parameters, the background initiating rate is estimated to be $\mu_0 = 1.229 \times 10^{-11}$. In calculating μ_0, it is assumed that the birth-death parameters are not dose dependent. This assumption appears reasonable in view of the fact that VC acts as a pure initiator [4, 9].

Dose-Response Model

In order to use the hazard function in Eq. 12 to predict the risk of liver cancer due to VC exposure, the relationship between dose and parameter values must be determined. In their study on VC-induced foci, Laib et al. [9] suggested that, in hepatocytes, the compound behaves like an "incomplete" carcinogen with only initiating properties. Therefore, it seems reasonable to assume that only μ_1 is dose dependent. In this paper, it is assumed that μ_1 is linearly proportional to dose metabolized; i.e., $\mu_1(d) = \mu_0 + cd$, where d is a VC-metabolized dose in μg, and c is a positive coefficient for d. This assumption is supported by the observation in Chen and Blancato [4] that tumor response is linearly related to the metabolized dose, irrespective of route of exposure. Table 4 gives the steady-state relationship between ppm exposure and mg metabolite per day for animals exposed to VC via inhalation for 8 hours per day, 5 days per week. These values are calculated using a physiologically based pharmacokinetic model [4].

Based on the fact that exposure to 2,000 ppm VC via inhalation, for 8 hours per day, 5 days per week, yields 12.64 mg per day of metabolites; the initiation rate due to 1 mg per day of metabolites would be 9.784×10^{-9} $(1.236 \times 10^{-7}/12.64)$. In general, the initiation rate is

$$\mu_1(d) = 1.229 \times 10^{-11} + 9.784 \times 10^{-9}d,$$

where d is a VC-metabolized dose in μg per day.

Table 4. Amount of total metabolites in rats exposed to VC via inhalation for 8 hours per day, 5 days per week

Exposure (ppm)	Total metabolites (mg/day)
10	0.28
40	1.06
70	1.80
100	2.42
150	3.44
500	8.78
2,000	12.64

SOURCE: Chen and Blancato [4].

Table 5 gives observed and predicted hepatocarcinoma incidences in female F344 rats exposed to 100 ppm VC by inhalation for 6 hours per day, 5 days per week, in studies from 0 to 24 months [7]. A data point in Table 5 is used to estimate parameter μ_2, and the remaining data are then used to test the adequacy of the model.

In using data in Table 5, it is assumed that Wistar rats and Fischer 344 rats respond to VC with respect to tumor formation in an identical fashion. The appropriateness of this assumption is in doubt because animals in the 6-12 month exposure group (Table 5) showed a significantly ($p < 0.005$, one-tailed test) higher tumor incidence (6/52) than the control group (1/112). This observation contradicts the Laib et al. [8, 9] observation (data not available) that adult animals exposed to VC did not show significant tumor increases. However, the implication of the tumor response observed in the 6-12 month group is ambiguous because another group of animals (6-18 month group) that had a duration of exposure two times longer than the 6-12 month group showed a significantly lower tumor incidence (1/54).

Table 5. Observed and predicted number of rats with hepatocarcinomas. Rats were exposed to 100 ppm VC via inhalation for 6 hours per day, 5 days per week, for different durations and ages of exposures[a]

Age and duration of exposure[b]	No. of rats	No. of rats with tumors	
		Observed	Predicted[c]
Control	112	1	0.0
0-6	76	3	4.6
0-12	55	4	6.5
0-18	55	8	8.4
0-24	55	9	[d]
6-12	52	6[e]	3.3
6-18	54	1[e]	5.3
12-18	51	0	1.9
16-24	49	0	1.1
18-24	53	1	0.7

[a] All animals were followed to death or sacrificed at 24 months.
[b] Exposures are per month.
[c] Calculated by probability of tumor, using the hazard function derived previously.
[d] Data used to estimate μ_2.
[e] These groups show numerically inconsistent responses.

SOURCE: Drew et al. [7].

The initiation rate corresponding to the experimental condition in Table 5 is calculated to be,

$$\mu_1 = 1.229 \times 10^{-11} + 9.784 \times 10^{-9} \times 1.82$$
$$= 1.781 \times 10^{-8},$$

where 1.82 µg per day is the estimated metabolized dose at the given experimental condition.

Using the tumor response (9/55) from the group of animals that were exposed to VC for an entire lifetime (0-24 months), μ_2 is estimated to be 1.486×10^{-8}. The predicted number of animals with tumors, along with the observed cases, are given in Table 5. Considering that only one data point is used to estimate μ_2 and the great variability of data, the model appears to fit the data adequately (see the footnote in Table 5).

This example demonstrates the usefulness of foci data for estimating parameters in the MVK model, under the assumption that foci are precursors of malignant tumors. Although the model appears to fit tumor dose-response data adequately, it by no means establishes that these foci are precursors of tumors. While the model supports the hypothesis that foci are preneoplastic, final judgement must come from biological inference. Another problem brought out by this example is that liver angiosarcomas that are known to be induced by VC are not considered because data on preneoplastic entities leading to angiosarcomas are not available. This problem points out the difficulty of constructing a meaningful BBDR model for the purpose of predicting cancer risk to humans. The difficulty comes from the fact that most of the experiments are not conducted for the purpose of dose-response modeling.

Example 2: B[a]P Skin-painting Study

Data Base

Tables 6 and 7 give the number of skin tumors per animal, in control and treated animals, that were exposed (by skin painting) to a single dose (128 µg) of B[a]P at 56 days of age and followed by 5 µg of TPA 3 times per week. These data are reconstructed from graphs in Burns et al. [3].

Parameter Estimation

According to Burns et al. [3], the area of exposed skin contains about 3×10^6 normal cells. Eq. 13 and Eq. 14 were used to estimate parameters for Model 2, using data in Tables 6 and 7 and the assumption that a papilloma contains at least 10^5 cells.

Table 6. Observed and predicted number of tumors per mouse in control animals

| Days after exposure began[a] | Tumors per mouse | | | |
| | Papillomas | | Carcinomas | |
	Observed	Predicted	Observed	Predicted
100	0	0.01	0	0
150	0.04	0.04	0	0
200	0.06	0.08	0	0
250	0.08	0.12	0	0
300	0.10	0.18	0	0.01
350	0.16	0.24	0.02	0.02
400	0.20	0.32	0.03	0.03
450	0.34	0.40	0.04	0.04
500	0.50	0.50	0.04	0.05

a Mice were exposed at 56 days of age.

SOURCE: Reconstructed from Figure 2; Burns et al. [3].

Table 7. Observed and predicted number of tumors for mice exposed to a single dose of B[a]P (128 µg) by skin painting, followed by 5 µg of TPA 3 times per week[a]

| Days after exposure began | Tumors per mouse | | | |
| | Papillomas | | Carcinomas | |
	Observed	Predicted	Observed	Predicted
50	0.3	0.70	0	0.00
100	1.2	1.58	0	0.04
150	2.4	2.45	0	0.10
200	3.1	3.29	0	0.19
250	3.7	4.12	0	0.31
300	5.0	4.94	0.1	0.45
350	6.3	5.75	0.3	0.62
400	6.3	6.54	0.3	0.81
450	6.8	7.31	1.0	1.03
500	9.2	8.08	1.6	1.27

a Mice were exposed at 56 days of age.

SOURCE: Reconstructed from Figure 3; Burns et al. [3].

In the calculations, it is assumed that the initiation rate is linearly dependent on the dose of B[a]P; i.e., $\mu_1(d) = \mu_0 + \mu_1 d$, where d is B[a]P dose in µg. The number of normal cells is assumed to be constant and is equal to 3×10^6. The resultant least square estimates of parameters (per cell per day) are as follows:

$\mu_0 = 1.53 \times 10^{-12}$, $\mu_1 = 4.68 \times 10^{-11}$ per µg B[a]P, $a_1 = 6.25 \times 10^{-4}$,
$b = 6.91 \times 10^{-2}$, and $d = 1.70 \times 10^{-3}$.

It should be noted that a_1, the transition rate from a papilloma to a carcinoma, and other parameters are estimated simultaneously from the same set of data. This estimating procedure differs from that used in the VC example where μ_2 is estimated separately from the tumor-response data after all other parameters are estimated on the basis of foci data. It should also be noted that a_1 in this example represents the mean transition rate from a papilloma to a carcinoma, not the transition rate from an I-cell to a cancer cell, as it does in Example 1.

Evaluation of the Model

Using the parameters obtained above, the number of papillomas and carcinomas per animal can be calculated under any dosing condition. Tables 6 and 7 give the numbers of observed and predicted papillomas and carcinomas per animal, which appear to fit these data adequately. Since the data in Tables 6 and 7 are used to estimate parameters, it is desirable to evaluate the model against other data sets. As shown in Table 8, the model also appears to

Table 8. Observed and predicted number of papillomas for mice exposed to a single dose of B[a]P (32 µg), followed by 5 µg TPA 3 times per week

Days after exposure	Papillomas per mouse	
	Observed	Predicted
50	0.20	0.18
100	0.55	0.41
150	0.70	0.64
200	0.85	0.88
250	0.98	1.12
300	1.29	1.37
350	1.30	1.62

SOURCE: Reconstructed from Figure 5; Burns and Albert [2].

adequately predict skin papillomas per animal under different dosing conditions. Table 8 compares only observed and predicted numbers of papillomas per animal up to 350 days because these are the only data available to us.

This example demonstrates that, if data are available, one can model tumor growth on the basis of preneoplastic entities (e.g., papillomas), rather than on the basis of individual cells. The advantage of modeling on the basis of preneoplastic entities alone is that cell independence need not be assumed. However, in this example, we have to rely on a mathematical formula that involves individual cells to estimate the formation rate of papillomas because such information is not available to us.

4. Discussion

The ability to model at the cellular level is important for reducing uncertainties associated with low-dose extrapolation because it is statistically more powerful to work with cells or foci than with whole animals. The two examples presented in this paper demonstrate that various preneoplastic entities (e.g., foci and nodules), can be used as a basis for estimating parameters (as in Example 1) and for modeling (as in Example 2). Ideally, a cancer bioassay should include a detailed study of the growth kinetics of preneoplastic entities such as foci and nodules. Useful data include the number and size distribution of these preneoplastic entities over time.

In this paper, we have demonstrated that use of the approximate form of the MVK model for dose-response modeling may not be appropriate. The use of the approximate MVK model for dose-response modeling could result in (1) misinterpretation of the biological implication of an observed tumor incidence and (2) erroneous estimation of parameters in the model. Moolgavkar et al. [14] also noted that the adequacy of the approximation critically depends on the birth rate, b, for an I-cell. Thus, the use of the approximate hazard function defeats the original intention to study the dose-response relationship by incorporating biological information into the dose-response model (i.e., in this case, clonal growth).

The VC example points out the difficulty of obtaining sufficient data for constructing a BBDR model. It is well known that VC induces not only carcinomas but also angiosarcomas in liver. Although Laib et al. [9] demonstrate a high correlation between liver foci area and angiosarcomas, the BBDR model constructed on the basis of hepatocytes and foci cannot be used to predict incidence of angiosarcomas.

Finally, if certain information about dynamics of foci and/or nodules is available, Model 2 offers an attractive approach for constructing a BBDR model without having to make assumptions that may violate biological reality (e.g., cell independence). It is our hope that such data will become more readily available in the near future.

Acknowledgements

We are grateful to Professor Grace Yang of the University of Maryland for her critical review and advice during the preparation of this paper. We also appreciate our discussion with Dr. David Reese, of the USEPA, Office of Health and Environmental Assessment, about some biological concepts involved in the paper.

174

References

1. Altman, P., Dittmer, D. Biology Data Book, 2nd edition, Vol. 1, Federation of American Societies for Experimental Biology, Bethesda, MD. (1972).

2. Burns, F., Albert, R. Mouse skin papillomas as early stage of carcinogenesis. J. Amer. College Toxicol. 1:29-45 (1982).

3. Burns, F., Albert, R., Altshuler, B., and Morris, E. Approach to risk assessment for genotoxic carcinogens based on data from mouse skin initiation-promotion model. Environ. Health Perspect. 50:309-320 (1983).

4. Chen, C., Blancato, J. Incorporation of biological information in cancer risk assessment. Example -- vinyl chloride. To appear in Toxicology and Industrial Health (in press).

5. Chiang, C. Introduction to stochastic processes in biostatistics, John Wiley & Sons, Inc., NY. (1968).

6. Dewanji, A., Venzon, D., and Moolgavkar, S. A stochastic two-stage model for cancer risk assessment: II. The number and size of premalignant clones. Risk Analysis 9(2):179-189 (1989).

7. Drew, R., Boorman, G., Haseman, J., McConnell, E., Busey, W., and Moore, J. The effect of age and exposure duration on cancer induction by a known carcinogen in rats, mice, and hamsters. Toxicol. Appl. Pharmacol. 58:120-130 (1983).

8. Laib, R., Klein, K., and Bolt, H. The rat liver foci bioassay: I. Age-dependence of induction by vinyl chloride of the ATPase-deficient foci. Carcinogenesis 6(1):65-68 (1985).

9. Laib, R., Pellio, T., Wunschel, U., Zimmermann, N., and Bolt, H. The rats liver foci bioassay: II. Investigation on the dose-dependent induction of ATPase-deficient foci by vinyl chloride at very low doses. Carcinogenesis 6(1):69-72 (1985).

10. Laib, R., Stockle, G., Bolt, H., and Kunz, W. Vinyl chloride and trichloroethylene: Comparison of alkylating effects of metabolites and induction of preneoplastic enzyme deficiencies in rat liver. J. Cancer Res. Clin. Oncol. 94:139-147 (1979).

11. Maltoni, C., Lefemine, G., Ciliberti, A., Cotti, G., and Carretti, D. Carcinogenicity bioassays of vinyl chloride monomer: A model of risk assessment on an experimental basis. Environ. Health Perspect. 41:3-29 (1981).

12.   Moolgavkar, S., Venzon, D. Two-event models for carcinogenesis: Incidence curves for childhood and adult tumors. Math. Biosciences 47:55-77 (1979).

13.   Moolgavkar, S., Knudson, A. Mutation and cancer: a model for human carcinogenesis. J. Natl. Cancer Inst. 66:1037-1052 (1981).

14.   Moolgavkar, S., Dewanji, A., and Venzon, D. A stochastic two-stage model for cancer risk assessment: I. The hazard function and the probability of tumor. Risk Analysis 8:383-392 (1989).

15.   Pugh, T., King, J., Koen, H., Nychka, D., Chover, J., Wahba, G., He, Y., and Goldfarb, S. Reliable stereological method for estimating the number of microscopic hepatocellular foci from their transections. Cancer Research 43:1261-1268 (1983).

16.   Serio, G. Two-stage stochastic model for carcinogenesis with time-dependent parameters. Statist. Prob. Lett. 2:95-103 (1984).

17.   Tan, W. A stochastic Gompertz birth-death process. Statist. Prob. Lett. 4:25-28 (1986).

18.   Tan, W., Gastardo, M. On the assessment of effects of environmental agents on cancer tumor development by a two-stage model of carcinogenesis. Math. Biosciences 73:143-155 (1985).

19.   Tan, W., Brown, C. A nonhomogeneous two-stage model of carcinogenesis. Math. Model. 9(8):631-642 (1987).

20.   Thorslund, T., Brown, C., and Charnley, G. Biologically motivated cancer risk models. Risk Analysis 7:109-119 (1987).

21.   Tucker, H. A stochastic model for a two-stage theory of carcinogenesis. Proceedings for the Fourth Berkeley Symposium, Berkeley, CA 4:387-403 (1961).

TRANSFORMATION AND GROWTH OF LUNG ADENOMAS IN MICE EXPOSED TO URETHANE

Stella Grosser

and

Alice S. Whittemore

Abstract Murine lung adenomas are benign tumors that may progress to malignant carcinoma. To gain insight into the mechanisms of urethane-induced lung carcinogenesis, we compare results from two experiments, each involving the addition of urethane to the drinking water of mice. In the first (the chronic experiment), animals were ɪosed continuously throughout the study. In the second (the acute experiment), animals were dosed only for a single two week period. In both experiments, animals were sacrificed at two week intervals from start of exposure. At each sacrifice all visible lung adenomas were counted and measured for size. We use the data from these experiments to test two hypotheses about the effects of chronic urethane exposure on adenoma transformation and growth. The *transformation independence* hypothesis postulates that each two-week exposure to urethane induces adenomas independently of all other such exposures. This hypothesis would hold if each adenoma resulted from a single transforming event, and if urethane had no effect on the proliferation rate of normal stem cells. The *growth independence* hypothesis postulates that adenoma growth is independent of urethane exposure. This hypothesis would hold if urethane had no effect on the proliferation rate of transformed adenoma-progenitor cells. We use the maximum likelihood method to test these hypotheses, and find that both are rejected by the data. Using the Double Poisson distribution [3], we also examine inter-animal dispersion in adenoma counts, and the consequences of overdispersion for hypothesis testing. Although the severe overdispersion found in these data substantially reduced the significance of the likelihood ratio statistic for the transformation independence hypothesis, the hypothesis still was rejected. Overdispersion had little effect on the likelihood ratio statistic that rejected the growth independence hypotheses. Reasons for rejection of the two hypotheses are discussed.

Keywords: adenomas; carcinogenesis; cell transformation; clonal growth; Double Poisson distribution; EM algorithm; maximum likelihood; urethane

Acknowledgements. This research was supported by NIH grant CA47448 and by a grant to SIMS from the Environmental Protection Agency. The authors are grateful to Dr. Ellen J. O'Flaherty, who conducted the experiments and suggested the hypotheses, and to Dr. Michael L. Dourson, who helped conduct the experiments.

1. INTRODUCTION.

Urethane (ethyl carbamate) is a rodent carcinogen, causing lung adenomas and adenocarcinomas in mice. The adenomas are benign tumors that may undergo further transition to malignant carcinoma. We shall think of them as clones of cells representing an intermediate stage in a multistage carcinogenic process. We seek evidence concerning the event(s) that transform a normal cell into one capable of generating an adenoma, and the effect of urethane on adenoma proliferation rates.

Insight into mechanisms for transformation and growth of urethane-induced adenomas can be gained from study of adenoma numbers and adenoma size distributions resulting from chronic and acute exposures. Here we compare results from two experiments conducted by Ellen J. O'Flaherty at the University of Cincinnati, each involving the addition of urethane to the drinking water of mice. In the first (the chronic experiment), animals were dosed continuously throughout their lives. In the second (the acute experiment), animals were dosed only for a single two week period, starting at age 41-45 days. In both experiments, animals were sacrificed at two week intervals from start of exposure. At each sacrifice all visible lung adenomas were counted and virtually all were measured for size. We shall use the data from these experiments to test two hypotheses about the effects of chronic exposure.

The first hypothesis postulates that each two-week exposure to urethane induces adenomas independently of all other such exposures. According to this hypothesis, the number of adenomas present in the lungs of a chronically exposed mouse is the number that would have been present had the animal received only a single two-week exposure, multiplied by the number of such exposures the animal actually received. Because this hypothesis states that each urethane exposure transforms stem cells independently of all other exposures, we shall call it the *transformation independence* hypothesis.

The second hypothesis postulates that adenoma growth is independent of urethane exposure. To test this hypothesis, we shall regard the distribution of adenoma sizes in a chronically exposed mouse as a mixture of size distributions, each corresponding to adenomas induced in a preceding two-week period. The second hypothesis states that the size distribution of adenomas induced in an earlier exposure period equals the distribution that would have resulted from a single two-week exposure at that time. Because this hypothesis states that adenoma growth rates are independent of urethane exposure, we shall call it the *growth independence* hypothesis.

2. EXPERIMENTAL METHODS

In both experiments urethane was added to the drinking water of outbred Swiss mice at a concentration of .05% by weight. Ten animals were killed at each of eight two-week intervals after start of exposure at age 41-45 days. The protocols for acute and chronic experiments differed only with respect to duration of urethane exposure. The excised lungs were expanded under 20 cm fluid pressure with Tellyesniczky's fixative and stored in fixative for at least seven days. After this time, the adenomas were easily seen as opaque nodules surrounded by translucent lung tissue. All adenomas visible in the lungs of each animal were counted, and their diameters were measured microscopically by cytometer. Diameters were classified into 10 size categories ranging from less than .2 mm to more than 1 mm. Further details of experimental methods can be found in [2,8].

3. THEORY

We formulate the basic model in discrete time, dividing the 16 week study period into eight two-week periods, denoted $t = 1, ..., 8$. We assume that an adenoma is the clone of a single transformed cell. We also assume that the probability of cell transformation and the stochastic laws governing growth of clones of transformed cells do not depend on the age of the animal when these events occur. Let g_u denote the expected number of epithelial stem cells transformed in period u, $u = 1, ..., 8$. We assume that no cells are transformed after period one in the acute experiment. This assumption is reasonable since urethane is completely cleared from an animal within 24 hours of administration [6,9]. Using the superscripts a and c to distinguish acute and chronic experiments, we write this assumption as

$$g_u^a = 0 \quad u = 2, ..., 8. \tag{3.1}$$

Also, since acute and chronic experiments had identical protocols in period one, the expected numbers of cells transformed in this period are the same:

$$g_1^a = g_1^c \equiv g_1. \tag{3.2}$$

The transformation independence (TI) hypothesis states that the expected number of cells transformed in each period of the chronic experiment equals the number transformed in period one:

$$TI: \quad g_u^c = g_1 \quad u = 2, ..., 8. \tag{3.3}$$

To describe the growth independence hypothesis, let $f(s, v)$ denote the probability that an adenoma has size s at the end of v periods after transformation, with $0 \leq f(s, v) \leq 1$, $s = 1, \ldots, 10$; $v = 0, 1, \ldots, 7$. Since no adenomas were observed before period three in either experiment, and since adenomas must reach size category one to be visible, we assume that growth to visibility requires at least two periods. That is,

$$f^a(s, v) = f^c(s, v) = 0 \quad s = 1, ..., 10; \quad v = 0, 1. \tag{3.4}$$

We also assume that all adenomas in the acute experiment are visible at the end of seven periods after transformation (Dourson and O'Flaherty,1982), so that

$$f^a(\cdot, 7) \equiv \sum_{s=1}^{10} f^a(s, 7) = 1. \tag{3.5}$$

(Throughout the paper we use the '·' notation to indicate summation over the missing index.) The growth independence (GI) hypothesis states that adenoma size probabilities among chronically exposed mice equal those among mice in the acute experiment:

$$GI: \quad f^c(s, v) = f^a(s, v) \equiv f(s, v), \quad s = 1, ..., 10; \quad v = 2, ..., 7. \tag{3.6}$$

We shall use the maximum likelihood method to test the transformation and growth independence hypotheses, using the distributions of adenoma numbers and adenoma sizes observed in the two experiments. To do so, we let $n(s, t, u)$ denote the number of adenomas of size s in a mouse at the end of period t that arise from cells transformed in period $u \leq t - 2$. We treat $n(s, t, u)$ as a random variable with mean $\mu(s, t, u) = g_u f(s, t - u)$. In the chronic experiment, we do not know when adenoma progenitor cells were transformed, so we do not observe the counts $n^c(s, t, u)$. Instead we observe the total numbers $n^c(s, t) = \sum_{u=1}^{t-2} n^c(s, t, u)$ of adenomas of a given size s, having mean $\mu^c(s,t) = \Sigma_{u=1}^{t-2} g_u^c f^c(s,t-u)$. In the acute experiment no transformation occurred after period one, so $n^a(s,t) = n^a(s,t,1)$ with mean $\mu^a(s,t) = g_1 f^a(s,t-1)$. Assumption (3.5) implies that $\mu^a(\cdot,8) = g_1$. The basic model is thus

$$\mu^a(s,t) = g_1 f^a(s,t-1), \quad t = 3,\ldots,8;$$

$$M_B: \quad \mu^a(\cdot,8) = g_1;$$

$$\mu^c(s,t) = \sum_{u=1}^{t-2} g_u^c f^c(s,t-u), \quad s = 1,\ldots,10; \quad t = 3,\ldots,8. \tag{3.7}$$

By substituting (3.3) into the third equation in (3.7) we obtain the transformation independence model

$$\mu^a(s,t) = g_1 f^a(s,t-1), \quad t = 3,\ldots,8;$$

$$M_{TI}: \quad \mu^a(\cdot,8) = g_1;$$

$$\mu^c(s,t) = g_1 \sum_{u=1}^{t-2} f^c(s,t-u), \quad s = 1,...,10; \quad t = 3,\ldots 8. \tag{3.8}$$

Similarly, substitution of (3.6) into (3.7) gives the growth independence model

$$\mu^a(s,t) = g_1 f(s,t-1), \quad t = 3,\ldots,8;$$

$$M_{GI}: \quad \mu^a(\cdot,8) = g_1;$$

$$\mu^c(s,t) = \sum_{u=1}^{t-2} g_u^c f(s,t-u), \quad s = 1,...,10; \quad t = 3,\ldots 8. \tag{3.9}$$

We first assume that the total number $n(\cdot,t)$ of adenomas observed in a mouse at the end of period t has a Poisson distribution with parameter $\mu(\cdot,t)$:

$$n(\cdot,t) \sim F_P[n(\cdot,t),\mu(\cdot,t)], \qquad t = 3,\ldots,8, \tag{3.10a}$$

where μ is determined by (3.7),(3.8) or (3.9), and

$$F_P[n,\mu] = \mu^n e^{-\mu}/n!, \quad n = 0,1,\ldots \tag{3.10b}$$

Adenoma counts in different mice are assumed independent. We also assume that, conditional on a mouse's total adenoma count $n(\cdot,t)$, its adenoma size distribution is multinomial with parameters $n(\cdot,t)$ and

$$\pi_t = (\pi_{1t},\ldots,\pi_{10.t}) = (\mu(1,t),\ldots,\mu(10,t))/\mu(\cdot,t), \quad t = 3,\ldots,8. \tag{3.11}$$

That is, an animal's observed adenoma sizes, given its total adenoma count, have the multinomial distribution

$$F_M\left[n(1,t),\ldots,n(10,t);\pi_t\right] = \frac{n(\cdot,t)!}{\prod_{s=1}^{10} n(s,t)!} \prod_{s=1}^{10} \pi_{s t}^{n(s,t)}. \qquad (3.12)$$

With these assumptions we can write the likelihood for all data from both acute and chronic experiments as a Poisson-based function L_P of the unknown mean adenoma counts $\mu(\cdot,3),\ldots,\mu(\cdot,8)$ times a multinomial-based function L_M of the unknown size probability vectors $\pi_3,\ldots,\pi_8$ (see Appendix).

We shall evaluate the transformation and growth independence hypotheses via likelihood ratio tests. For example, to test the transformation independence hypothesis, we shall compare the maximum value of the likelihood function under the basic model M_B of (3.7) with the maximum value under the more restrictive model M_{TI} of (3.8). The comparison will be based on the likelihood ratio statistic

$$\lambda = 2log\left[\hat{L}(M_B)/\hat{L}(M_{TI})\right] \qquad (3.13)$$

where $\hat{L}$ denotes the maximized value of the likelihood function L. When the transformation independence hypothesis is true, λ has approximately a chi-squared distribution, so larger values of λ than expected from such a distribution suggest rejection of the hypothesis. Similarly, we shall test the growth independence hypothesis by comparing the maximized loglikelihood under M_B to that of the model M_{GI} of (3.9).

4. RESULTS

As described in the Appendix, the maximum likelihood estimates (MLE's) for the 2X6 = 12 expected adenoma counts $\mu^a(\cdot,t), \mu^c(\cdot,t), t = 3,\ldots,8$ and the 2X6=12 size probability vectors $\pi_t^a, \pi_t^c, t = 3,\ldots,8$ in the basic model M_B are the observed mean counts

$$\hat{\mu}(\cdot,t) = n_\bullet(\cdot,t)/r_t, \qquad (4.1a)$$

$$\hat{\pi}_t = \left(n_\bullet(1,t),\ldots,n_\bullet(10,t)\right)/n_\bullet(\cdot,t), \quad t = 3,\ldots,8. \qquad (4.1b)$$

Here we have omitted the superscripts a and c, and $n_\bullet(s,t)$ denotes the total number of adenomas among all r_t mice killed at time t. Note that there are a total of

182

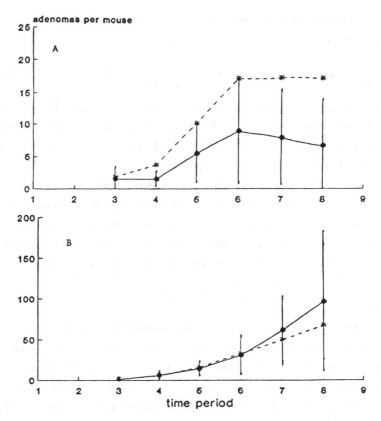

Figure 1. Observed mean adenoma counts ($\bullet$————$\bullet$) and counts predicted by the transformation independence model M_{TI}($* - - - *$) in A) acute and B) chronic experiments.

$12 + (12X9) = 120$ parameters in the basic model. Figure 1 shows the mean counts $\hat{\mu}(\cdot, t)$ for acute and chronic experiments.

4.1 Transformation Independence.

The model M_{TI} also involves 120 transformation and growth parameters, namely the 12 parameters $g_1, f^a(\cdot, v), f^c(\cdot, v)$, $v = 2, \ldots, 7$, with $f^a(\cdot, 7) = 1$, and the 12x9 parameters in the size probability vectors π_t^a, π_t^c $t = 3, \ldots, 8$. Thus M_{TI} requires the same number of adjustable parameters as does the basic model M_B. Therefore we cannot test the transformation independence hypothesis without further assumptions about the growth probabilities $f^c(\cdot, v)$ in the chronic experiment. The need for such assumptions can be seen heuristically by noting that without them, discrepant numbers of adenomas in acute and chronic experiments could arise either from discrepant transformation probabilities ($g_u^c \neq g_1$), or from discrepant probabilities that transformed cells become visible adenomas ($f^c(\cdot, v) \neq f^a(\cdot, v)$). The first discrepancy violates transformation independence. The second discrepancy implies that urethane affects early previsible clonal growth of transformed cells. Additional data are needed to distinguish these possibilities.

We deal with this problem by assuming that rates of adenoma growth to visibility are equal in acute and chronic experiments. In symbols,

$$f^c(\cdot, v) = f^a(\cdot, v) \equiv f(\cdot, v) \quad v = 2, \ldots, 7. \tag{4.2}$$

The transformation independence model M_{TI} is now specified by (3.8) and (4.2). This model puts no constraints on the individual size probabilities π_t^c and π_t^a of (3.11). Therefore the multinomial likelihood components L_M under M_{TI} and M_B are identical. Hence their maximized values are equal, and the likelihood ratio statistic depends only on the Poisson component L_P of the likelihood function. That is,

$$\lambda = 2log\big[\hat{L}(M_B)/\hat{L}(M_{TI})\big] = 2log\big[\hat{L}_P(M_B)/\hat{L}_P(M_{TI})\big], \tag{4.3}$$

showing that the observed adenoma size distributions contribute no information toward testing the transformation independence hypothesis.

The Poisson likelihood L_P under the basic model M_B requires the 12 parameters $\mu^a(\cdot, t), \mu^c(\cdot, t), t = 3, \ldots, 8$. By constrast, M_{TI} of (3.8) and (4.2) requires only the six parameters $\gamma = g_1$ and $\Phi = [f(\cdot, 2), \ldots, f(\cdot, 6)]$. MLE's for (γ, Φ) under M_{TI} are not available in closed form. To obtain them, we used the EM algorithm [1], described in the Appendix. The results, shown on the top of Figure 2, indicate that the data reject the transformation independence hypothesis. The likelihood ratio statistic λ of (4.3) is 380.0, very large for a chi-squared statistic on 12-6=6 degrees of freedom ($p < .001$). Furthermore, Figure 1 shows that the adenoma counts observed in the acute experiment are smaller than those predicted by M_{TI}, while observed counts in the chronic experiment are larger than predicted. Therefore either a) the number of cells transformed in one period of the chronic experiment

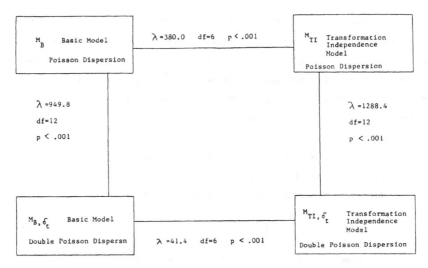

Figure 2. Nested models used to test the transformation independence hypothesis. M_B denotes the basic model (3.7) and M_{TI} denotes the transformation independence model obtained under the additional constraints (3.3) and (4.2). The subscript δ_t denotes a Double Poisson model in which the dispersion parameters δ are allowed to vary with experiment and time. Omission of this subscript denotes an ordinary Poisson model $(\delta_t \equiv 1)$. Lines connecting a pair of nested models are shown with the likelihood ratio statistic λ comparing the two models.

exceeds the number transformed in a single acute exposure $(g_u^c > g_1, u = 2, \ldots, 6)$, or b) urethane increases the chance that a transformed cell generates a visible clone in $v \leq 7$ periods $(f^c(\cdot, v) > f^a(\cdot, v), \quad v = 2, \ldots, 7.)$

4.2 Growth Independence.

The likelihood function under model M_{GI} of (3.9) depends on 6+5+54=65 parameters: the vector $\gamma = (g_1, \ldots, g_6)$ of six transformation parameters, the vector $\Phi = (f(\cdot, 2), \ldots, f(\cdot, 6))$ of five visibility probabilities, and the 6X10 matrix $\Psi = (\psi_{v\cdot})$ of 54 size probabilities $\psi_{v\cdot} = f(s, v)/f(\cdot, v), \quad s = 1, \ldots, 10, \quad \Sigma_s \psi_{v\cdot} = 1, \quad v = 2, \ldots, 7$. Maximizing it requires an iterative procedure; we used the EM algorithm. The details are provided in the Appendix.

Figure 3 shows that the data reject the growth independence hypothesis. The likelihood ratio statistic for the pair (M_B, M_{GI}) is 111.0 on 120-65=55 degrees of freedom $(p < .001)$.

Such rejection could be due either to acceleration or to retardation of trans-

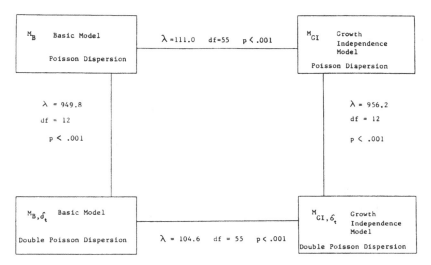

Figure 3. Nested models used to test the growth independence hypothesis. M_B denotes the basic model (3.7) and M_{GI} denotes the growth independence model (3.9). The subscript δ_t denotes a Double Poisson model in which the dispersion parameters δ are allowed to vary with experiment and time. Omission of this subscript denotes an ordinary Poisson model ($\delta_t \equiv 1$). Lines connecting a pair of nested models are shown with the likelihood ratio statistic λ comparing the two models.

formed cell proliferation rates in the chronic relative to the acute experiment. To distinguish these possibilities, we present in Figure 4 the observed adenoma size distributions at various times of the acute and chronic experiments and the distributions predicted under M_{GI}. The latter are given by

$$\hat{\pi}^a_{st} = \hat{f}(s, t-1)/\hat{f}(\cdot, t-1)$$

$$\hat{\pi}^c_{st} = \sum_{u=1}^{t-2} \hat{g}_u \hat{f}(s, t-u) / \sum_{u=1}^{t-2} \hat{g}_u \hat{f}(\cdot, t-u) \qquad s = 1, \ldots, 10; \quad t = 3, \ldots, 8, \qquad (4.4)$$

where $\hat{g}_u$ and $\hat{f}(s, t-u)$ are the MLE's under M_{GI}. Figure 4 shows that at early times the model predicts too few large adenomas in the acute experiment, and too many large adenomas in the chronic experiment. For late times the fit is better. The discrepancies suggest that some concomitant of chronic urethane exposure retards adenoma growth.

4.3 Dispersion in Adenoma Counts.

We have seen that both the transformation independence and the growth independence hypotheses are rejected when we assume a Poisson distribution for each

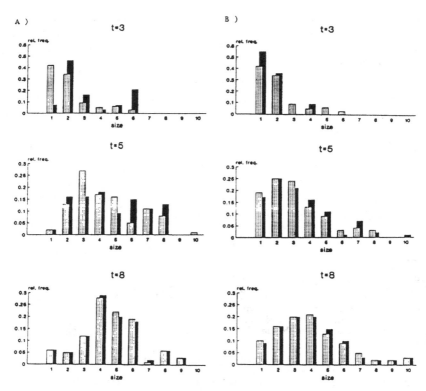

Figure 4. Observed adenoma size distributions (▨) and those predicted by the growth independence model M_{GI} (■) at times $t = 3,5$ and 8 in A) acute and B) chronic experiments. Observed and predicted size distributions are given, respectively, by text-equations (4.1b) and (4.4).

animal's adenoma count. However the data present an obstacle to this analysis. Specifically, the observed mouse-to-mouse variance in adenoma counts at a given time almost always exceeds the mean adenoma count at that time. Because a Poisson random variable has variance equal to its mean, this overdispersion makes the Poisson assumption suspect. If not accounted for in the analysis, the overdispersion might lead to incorrect rejection of the hypothesis under test. To account for it, we shall assume that the adenoma counts are distributed according to a Double Poisson distribution, as introduced by Efron [3].

A random variable X having a Double Poisson distribution with mean $\mu > 0$

and dispersion parameter $\delta > 0$ has the probability density function

$$P(X = n) = c(\mu, \delta)\delta^{-\frac{1}{2}}\mu^{n/\delta}n^{(\delta-1)n/\delta}e^{-[(\mu+(\delta-1)n)/\delta]}/n!, \quad n = 0, 1, \ldots.$$

Here $c(\mu, \delta)$ is a normalization constant chosen to make the distribution sum to one over the values $n = 0, 1, 2, \ldots$. The constant $c(\mu, \delta)$ is approximately one, so following Efron [3], we will work with the approximate distribution function

$$F_{DP}(n; \mu, \delta) = \delta^{-\frac{1}{2}}\mu^{n/\delta}n^{(\delta-1)n/\delta}e^{-[(\mu+(\delta-1)n)/\delta]}/n!, \quad n = 0, 1, \ldots \quad (4.5)$$

The Double Poisson distribution generalizes the Poisson distribution by introducing the extra dispersion parameter δ in addition to the mean parameter μ. The parameter δ measures overdispersion (or underdispersion) in adenoma counts from one mouse to another. When $\delta = 1$, the Double Poisson distribution reduces to the Poisson distribution (3.10b), for which the mean and variance of adenoma numbers are equal. When $\delta > 1$, the variance exceeds the mean, and the adenoma numbers are said to be overdispersed. When $\delta < 1$, the variance is less than the mean, indicating underdispersion. Indeed, δ is approximately the ratio of the distribution variance to the distribution mean [3]. (The negative binomial distribution also allows the introduction of overdispersion, but maximum likelihood estimates for its parameters are more difficult to obtain than those of the Double Poisson distribution.)

We shall replace (3.10) by the assumption that the total number $n(\cdot, t)$ of adenomas observed in a mouse at the end of period t has a Double Poisson distribution of the form (4.5) with parameters $\mu(\cdot, t)$ and δ_t, $t = 3, \ldots, 8$. Adenoma counts in different mice are again assumed independent. The resulting likelihood is given by Appendix equation (A.1), but with the Poisson factors $F_P[n(\cdot, t); \mu(\cdot, t)]$ replaced by the Double Poisson factors $F_{DP}[n(\cdot, t), \mu(\cdot, t), \delta_t]$, where F_{DP} is given by (4.5). Now the Poisson component of the likelihood depends not only on the unknown mean parameters, but also on the unknown dispersion parameters $\delta_3, \ldots, \delta_8$ in each experiment.

An attractive feature of Double Poisson regression models is the independence of estimates for their mean parameters from the form assumed for their dispersion parameters. In our application, this independence implies that MLE's for the expected adenoma counts $\mu(\cdot, t)$ and size probability vectors π_t are unaffected by the way the dispersion parameters are modelled. Therefore the estimated adenoma counts and size distributions previously obtained for the Poisson-based models also apply to the more general Double-Poisson-based models. However values for the maximized likelihoods and for the likelihood ratio statistics are different. Moreover while estimates for the mean parameters are independent of the assumed form for the $\delta'_t s$, the reverse is not true. That is, the estimates $\hat{\delta}_t$ depend on the form assumed for the mean parameters. It is easy to check that the MLE's for the δ_t's under the Double Poisson models are:

$$\hat{\delta}_t = r_t^{-1} \sum_{i=1}^{r_t} D\big[n_i(\cdot, t), \hat{\mu}(\cdot, t)\big], \tag{4.6a}$$

where the $\hat{\mu}(\cdot, t)$ are the MLE's specified by the given model, and

$$D(n, \mu) = 2\big[n log(n/\mu) - (n - \mu)\big] \tag{4.6b}$$

is the Poisson deviance [5].

Once we have maximized the Poisson-based likelihoods using the methods described in the Appendix, it is easy to calculate maximized values for the Double Poisson likelihoods. These are given by Appendix equation (A.1) with the Poisson component replaced by its corresponding Double Poisson component, with the unknown mean parameters $\mu(\cdot, t)$ replaced by their Poisson MLE's, and with the unknown dispersion parameters δ_t replaced by (4.6).

The left panels of Figures 2 and 3 show the likelihood ratio statistic for comparing the basic Double Poisson model M_{B,δ_t} to its Poisson submodel M_B. As expected, the Poisson submodel is strongly rejected by the data, as indicated by the value $\lambda(M_{B,\delta_t}, M_B) = 949.8$ $(p < .001)$. The MLE's for $(\delta_3, \ldots, \delta_8)$ were $(2.1, 0.9, 3.6, 7.0, 6.5, 7.8)$ for the acute experiment and $(1.5, 3.1, 4.9, 16.3, 24.1, 52.0)$ for the chronic experiment. These values indicate overdispersion in the adenoma counts. The overdispersion increases with time, as do the counts themselves. Indeed a submodel for which the δ_t are constant over time was rejected $(\chi^2_{10} = 55.4; p < .001)$.

Note that because the multinomial component L_M is independent of the δ_t's, it is invariant with respect to the model for the δ_t, and does not contribute to likelihood ratio statistics comparing one such model to another. That is, the size distributions do not contribute to tests for overdispersion in the adenoma counts.

Figure 2 also shows the likelihood ratio statistic comparing the transformation independence model M_{TI,δ_t} to the basic model M_{B,δ_t} when the dispersion parameters are allowed to vary freely with experiment and time. Allowing for overdispersion decreases the likelihood ratio statistic from 380.0 to 41.4. Despite this substantial decrease however, the transformation hypothesis still is rejected ($p < .001$).

Figure 3 gives the corresponding comparison for the growth independence model M_{GI,δ_t}. Allowing for overdispersion only slightly depressed the likelihood ratio statistic comparing basic and growth independence models from its value under the Poisson distributional assumption.

In summary, the data from both acute and chronic experiments show marked overdispersion in the adenoma counts. However allowing for this overdispersion does not salvage either the transformation independence hypothesis or the growth independence hypothesis.

6. DISCUSSION

We have used the maximum likelihood method to test two hypotheses about the effects of chronic urethane exposure on the kinetics of adenogenesis in mice. We find that the data reject both the transformation independence hypothesis (stipulating additivity of successive exposure increments in transforming normal cells), and the growth independence hypothesis (stipulating no effect of chronic exposure on adenoma proliferation rates). Allowance for the severe mouse-to-mouse variation noted in the adenoma counts did not alter these conclusions.

Some limitations of the design and analysis must be considered in interpreting these results. First, the smallest (less than .2 mm) adenomas may have been incompletely noted. It is unlikely, however, that incomplete observation of small adenomas could account for the large observed differences in adenoma numbers that caused rejection of the transformation independence hypothesis, or for the differences in size distributions that caused rejection of the growth independence hypothesis.

Second, we have assumed that the probability of transformation and cell division depends only on the timing of urethane exposure, and not on the ages of the mice when these events occur. Since transformation occurred over a 16 week period in the chronic experiment and only within a two week period in the acute experiment, such age effects, if present, could explain the differences in adenoma sizes and numbers noted in the two experiments. The relative youth (about 23 weeks old) of the mice at the end of the experiments suggests that effects due to old age are unlikely. Nevertheless we cannot dismiss the possibility that age effects may account for the results. Additional experiments involving acute urethane exposure in periods $2, \ldots, 6$ are needed to distinguish exposure effects from age effects. The absence of age effects noted for benzpyrene-induced murine skin papillomas in such a series of staggered acute experiments [7] suggests that, if adenomas behave as papillomas, such effects may not be important here.

The transformation independence hypothesis was rejected because too many adenomas were found in the chronic experiment, compared to the numbers expected from the acute experiment. Although several explanations for this discrepancy come to mind, one seems most plausible. Chronic urethane exposure may increase the number of stem cells at risk of transformation, by increasing their net mitotic rate. There is evidence to support the possibility of such urethane-induced stem cell hyperplasia. Kauffman [4] counted the number of type-II alveolar cells in the lungs of mice after 1-16 weeks of exposure to urethane in drinking water. After an initial decrease in the first two week period, cell numbers increased. Indeed, Kauffman's plot of cell numbers versus time since start of exposure is roughly proportional to estimated mean numbers $\hat{g}_i^c$ of cells transformed in the present chronic experiment. This proportionality suggests that adenomas arise from type-II alveolar cells after one genetic event, and that rejection of the transformation independence hypothesis is due to alveolar cell hyperplasia in the chronic experiment.

Alternatively, the transformation independence hypothesis may have been rejected because urethane requires two or more genetic events to transform a normal stem cell to one capable of generating an adenoma. However available data [6,8] suggest a linear relationship between tumor numbers and estimated internal exposure to urethane. Such a linear relationship would imply that urethane affects only one genetic event in cell transformation.

The growth independence hypothesis was rejected because too few large adenomas were observed in the chronic experiment, compared to the numbers expected from the acute experiment. Possible explanations are: i) chronic urethane exposure may cause poor weight gain, which may inhibit adenoma growth; ii) the large numbers of adenomas in the lungs of chronically exposed mice may inhibit their growth (by, say, limiting the blood supply to each adenoma); iii) urethane may decrease the net proliferation rate of adenoma cells.

The first explanation seems unlikely, because the mice were exposed to urethane only for 16 weeks. Chronic urethane administration can result in debility with poor weight gain or even weight loss. However, at 0.05% urethane in the drinking water, such effects occur only after long-term exposure (six months or more). Mice in whom impaired growth has been observed have been found to have a large fraction of the total lung occupied by tumor tissue. It is reasonable to interpret impaired growth as an indirect consequence of severely impaired lung function. However exposed mice in the present acute and chronic experiments did not exhibit this degree of lung tissue involvement, nor was their growth rate noticeably different from that of control mice.

Further data are needed to distinguish the second and third explanations for rejection of the growth independence hypothesis. The direction in which the hypothesis was rejected indicates that chronic exposure retards rather than accelerates adenoma growth. Thus urethane, a potent initiator, does not seem to have promoting activity in this system.

Whatever be the explanations for the observed differences in transformation and growth of urethane-induced adenomas, the differences suggest caution in extrapolating to cancer risks from chronic exposures on the basis of risks estimated from acute experiments.

Although allowing for overdisperison in adenoma counts did not alter the above conclusions, it did substantially reduce the statistical significance of the transformation independence test. Tumor overdispersion is common in rodent carcinogenesis experiments. The present results indicate the importance of accounting for it when making inferences based on tumor numbers, because failure to do so may produce misleadingly narrow confidence intervals and type-I-error distortion in statistical testing.

REFERENCES

1. Dempster AP, Laird NM, Rubin DB. Maximum likelihood from incomplete data via the EM algorithm (with discussion). Journal of the Royal Statist. Soc. B 39: 1-38, 1977.

2. Dourson ML and O'Flaherty EJ. Relationship of lung adenoma prevelance and growth rate to acute urethan dose and target cell number. JNCI 69: 851-857, 1982.

3. Efron B. Double exponential families and their use in generalized linear regression. J. Amer. Statist. Assoc. 81: 709-721, 1986.

4. Kauffman SL. Autoradiographic study of type II-cell hyperplasia in lungs of mice chronically exposed to urethane. Cell Tissue Kinet. 9:489-497, 1976.

5. McCullagh P and Nelder JA. Generalized Linear Models. Chapman and Hall, New York, 1983.

6. O'Flaherty EJ and Sichak SP. The kinetics of urethane elimination in the mouse. Toxicol. Appl. Pharmocol. 68: 354-358,1983.

7. Peto R, Roe FJC, Lee PN, Levy L, Clack J. Cancer and ageing in mice and men. Br. Journal of Cancer 32: 411-426, 1975.

8. Sichak SP and O'Flaherty EJ. Consideration of the mechanism of pulmonary adenogenesis in urethane-treated Swiss mice. Toxicol. Appl. Pharmocol 76: 397-402, 1984.

9. White M. Studies of the mechanism of induction of pulmonary adenomas in mice. In Proc. of Sixth Berkeley Symp. on Math. Stats. and Prob. L.M.leCam, J.Neyman, E.L.Scott,eds. Vol IV pp 287-307. UC Berkeley Press, Berkeley,CA 1972.

APPENDIX: MAXIMUM LIKELIHOOD METHODS

The basic model M_B of (3.7), together with the Poisson and multinomial assumptions (3.10)-(3.12), imply that the likelihood function for the data from acute and chronic experiments is the product of a Poisson component for the adenoma counts times a multinomial component for the adenoma sizes:

$$
\begin{aligned}
L\big(\mu^a\,(\cdot,t),\pi^a_t,\mu^c\,(\cdot,t),\pi^c_t \underline{i}gr\big) &= L_P\big(\mu^a\,(\cdot,t),\mu^c\,(\cdot,t),\big) \\
&\quad \times L_M\big(\pi^a_t,\pi^c_t\,;t=3,\ldots,8\big).
\end{aligned}
\tag{A.1a}
$$

Here

$$
\begin{aligned}
L_P\big(\mu^a\,(\cdot,t),\mu^c\,(\cdot,t)\big) &= \prod_{t=3}^{8}\prod_{i=1}^{r^a_t} F_P\big[n^a_i(\cdot,t);\mu^a(\ cdot,t),\big] \\
&\quad \times \prod_{t=3}^{8}\prod_{j=1}^{r^c_t} F_P\big[n^c_j(\cdot,t);\mu^c(\cdot,t)\big],
\end{aligned}
\tag{A.1b}
$$

$$
\begin{aligned}
L_M\big(\pi^a_t,\pi^c_t\,;t=3,\ldots,8\big) &= \prod_{t=3}^{8}\prod_{i=1}^{r^a_t} F_M\big[(n^a_i(1,t),\ldots,n^a_i(10,t));\pi^a_t\big] \\
&\quad \times \prod_{t=3}^{8}\prod_{j=1}^{r^c_t} F_M\big[(n^c_j(1,t),\ldots,n^c_j(10,t));\pi^c_t\big],
\end{aligned}
\tag{A.1c}
$$

F_P and F_M are given by (3.10) and (3.12) respectively, and r_t denotes the number of mice killed at the end of period t.

The likelihood function L involves 120 parameters $\mu^a(\cdot,t)$, $\mu^c(\cdot,t)$, π^a_t, π^c_t; $t=3,\ldots,8$, for which the observed mean adenoma counts and size distributions form a set of minimal sufficient statistics. The maximum likelihood estimates (MLE's) are thus

$$
\hat{\mu}(\cdot,t) = n_{\bullet}(\cdot,t)/r_t,
\tag{A.2a}
$$
$$
\hat{\pi}_t = \big(n_{\bullet}(1,t),\ldots,n_{\bullet}(10,t)\big)/n_{\bullet}(\cdot,t), \quad t=3,\ldots,8,
\tag{A.2b}
$$

where we have omitted the superscripts a and c. In this saturated model, MLE's for the expected adenoma counts μ are the mean adenoma numbers per mouse observed at a given time, and MLE's for the size proportions $\pi_t = (\pi_{1,t},\ldots,\pi_{10,t})$ are the observed size proportions.

The likelihood ratio test (4.3) for the transformation independence model M_{TI} depends only on maximized values of the Poisson likelihood component L_P under M_B and M_{TI}. Under M_B, L_P depends on the 12 parameters $\mu^a(\cdot,t),\mu^c(\cdot,t)$, $t=3,\ldots,8$. Under M_{TI} by contrast, L_P depends only on the six parameters $\gamma = g_1$ and $\Phi = [f(\cdot,2),\ldots,f(\cdot,6)]$. Since MLE's for these six parameters are not available in closed form, we use the EM algorithm to obtain them. To do so, we treat as missing data the unobserved counts $n^c(\cdot,t,u)$ of adenomas seen at time t and transformed in period $u \leq t-2$ of the chronic experiment. We call the counts $n(\cdot,t,u)$ the complete data, and the counts $n(s,t) = \sum_{u=1}^{t-2} n(\cdot,t,u)$ the observed data. (Notice that complete and observed data coincide for the acute experiment, because $n^a(s,t,u) = 0$ for $u \neq 1$.) The objective is to find (γ,Φ) to maximize the Poisson likelihood $L_P(\gamma,\Phi)$ given by (A.1b).

The EM algorithm consists of two steps which transform a current estimate (γ^m,Φ^m) to a new one $(\gamma^{m+1},\Phi^{m+1})$. The two steps are iterated until convergence.

In the E-step we compute $Q(\gamma, \Phi)$, the expectation of the complete data loglikelihood, conditional on the observed data and on the current estimate (γ^m, Φ^m). In the M-step we maximize Q with respect to (γ, Φ) to obtain the new estimate. The algorithm works because maximizing Q is simpler than maximizing L_P.

Under M_{TI} the complete data likelihood corresponding to $L_P(\gamma, \Phi)$ is

$$
\mathcal{L}_P(\gamma, \Phi) = \prod_{t=3}^{8} \prod_{i=1}^{r_t^a} F_P\left[n_i^a(\cdot, t); g_1 f(\cdot, t-1)\right]
$$

$$
\times \prod_{t=3}^{8} \prod_{u=1}^{t-2} \prod_{j=1}^{r_t^c} F_P\left[n_j^c(\cdot, t, u); g_u f(\cdot, t-u)\right],
$$

(A.3)

with $g_1 = g_2 = \ldots = g_6$. The E-step of the algorithm is easily accomplished because $\log \mathcal{L}_P$ is linear in the counts $n_j^c(\cdot, t, u)$, so that its expected value equals $\log \mathcal{L}_P$ with the $n_j^c(\cdot, t, u)$ replaced by their expected values $\eta_j(\cdot, t, u)$, conditional on the $n_j^c(\cdot, t)$ and on the current parameters estimates. These are given by

$$
\eta_j(\cdot, t, u) = n_j^c(\cdot, t)\left[\sum_{v=2}^{t-1} f^m(\cdot, v)\right]^{-1} f^m(\cdot, t-u), \quad u = 1, \ldots, t-2.
$$

(A.4)

Thus the E-step uses the current parameter estimates to partition the observed number $n_j^c(\cdot, t)$ of adenomas on the jth mouse at time t into $t-2$ subclasses of sizes $\eta_j(\cdot, t, u), u = 1, \ldots, t-2$, where the uth subclass consists of adenomas transformed in period u.

In the M-step we maximize (A.3) with $\eta_j(\cdot, t, u)$ replacing $n_j^c(\cdot, t, u)$. The MLE's satisfy

$$
\gamma^{m+1} = g_1^{m+1} = \frac{\sum_{t=3}^{8} \left[n_\bullet^a(\cdot, t) + \sum_{u=1}^{t-2} \eta_\bullet(\cdot, t, u)\right]}{\sum_{t=3}^{8} \left[r_t^a f^{m+1}(\cdot, t-1) + r_t^c \sum_{u=1}^{t-2} f^{m+1}(\cdot, t-u)\right]}, \quad (A.5a)
$$

$$
f^{m+1}(\cdot, v) = \frac{n_\bullet^a(\cdot, v+1) + \sum_{t=v+1}^{8} \eta_\bullet(\cdot, t, t-v)}{g_1^{m+1}(r_{v+1}^a + \sum_{t=v+1}^{8} r_t^c)}, \quad v = 2, \ldots, 7. \quad (A.5b)
$$

Equations (A.5) can be solved iteratively by setting the $f^{m+1}(\cdot, v)$ equal to some initial values, using them in (A.5a) to get values for g_1^{m+1}, using this in (A.5b) to get new values for the $f^{m+1}(\cdot, v)$, etc.

The EM algorithm also is needed to maximize the likelihood function L of (A.1) under the growth independence model M_{GI}. Using (3.9) and (3.12), we reparameterize L as

$$
L(\gamma, \Phi, \Psi) = L_P(\gamma, \Phi) \cdot L_M(\gamma, \Phi, \Psi), \quad (A.6a)
$$

where

$$
\gamma = (g_1, g_2^c, \ldots, g_6^c),
$$
$$
\Phi = (f(2, \cdot), \ldots, f(6, \cdot)),
$$

and

$$
\Psi = (\psi_{v\bullet}) \quad \text{with} \quad \psi_{v\bullet} = f(s, v)/f(\cdot, v), \quad s = 1, \ldots, 10, \quad v = 2, \ldots, 7. (A.6b)
$$

L involves six parameters in γ, plus 5 parameters in Φ, plus 6x10=66 parameters in the matrix Ψ, for a total of 65 parameters. Neither the Poisson component nor the multinomial component of L is saturated under M_{GI}, so that the likelihood ratio statistic cannot be reduced as in (4.3).

To maximize L with respect to (γ, Φ, Ψ), we let the complete data consist of all observed size-specific adenoma counts, plus the unobserved counts $n_j^c(s,t,u)$. The complete data likelihood is then

$$\mathcal{L}(\gamma, \Phi, \Psi) = \mathcal{L}_P(\gamma, \Phi) \times \mathcal{L}_M(\Psi), \qquad (A.7a)$$

where $\mathcal{L}_P$ is given by (A.3) and

$$\mathcal{L}_M(\Psi) = \prod_{t=3}^{8} \prod_{i=1}^{r_i^a} F_M\left[(n_i^a(1,t),\dots,n_i^a(10,t)); \Psi_{t-1}\right]$$

$$\times \prod_{t=3}^{8} \prod_{u=1}^{t-2} \prod_{j=1}^{r_i^c} F_M\left[(n_j^c(1,t,u),\dots,n_j^c(10,t,u)); \Psi_{t-u}\right], \qquad (A.7b)$$

with Ψ_ν denoting the νth row of Ψ, $\nu = 2,\dots,7$. The expected value of the complete data loglikelihood equals $log\mathcal{L}$ with the unobserved counts $n_j^c(s,t,u)$ replaced by their expected values $\eta_j(s,t,u)$. These are

$$\eta_j(s,t,u) = n_j^c(s,t)\left[\sum_{v=2}^{t-1} g_{t-v}^m f^m(s,v)\right]^{-1} g_u^m f^m(s,t-u), \quad u = 1,\dots,t-2. \quad (A.8)$$

Since the expected value of $log\mathcal{L}_P$ depends only on (γ, Φ) and the expected value of $log\mathcal{L}_M$ depends only on Ψ, we can maximize them separately. The MLE's for (γ, Φ) satisfy

$$g_1^{m+1} = \frac{\sum_{t=3}^{8}[n_\bullet^a(\cdot,t) + \eta_\bullet(\cdot,t,1)]}{\sum_{t=3}^{8}(r_t^a + r_t^c)f^{m+1}(\cdot,t-1)}; \qquad (A.9a)$$

$$g_u^{m+1} = \frac{\sum_{t=u+2}^{8} \eta_\bullet(\cdot,t,u)}{\sum_{t=u+2}^{8} r_t^c f^{m+1}(\cdot,t-u)} \quad u = 2,\dots,6; \qquad (A.9b)$$

$$f^{m+1}(\cdot,v) = \frac{n_\bullet^a(\cdot,v+1) + \sum_{t=v+1}^{8} \eta_\bullet(\cdot,t,t-v)}{r_{v+1}^a g_1^{m+1} + \sum_{t=v+1}^{8} r_t^c g_{t-v}^{m+1}}, \quad v = 2,\dots,7. \quad (A.9c)$$

Equations (A.9) can be solved iteratively by setting the $f^{m+1}(s,v)$ equal to some initial values, using them in (A.9a,b) to get values for the g_u^{m+1}, using these in (A.9c) to get new values for the $f^{m+1}(s,v)$, etc. MLE's for the entries of $\Psi = (\psi_{v\bullet})$ are

$$\psi_{v\bullet}^{m+1} = \frac{n_\bullet^a(s,v+1) + \sum_{t=v+1}^{8} \eta_\bullet(s,t,t-v)}{n_\bullet^a(\cdot,v+1) + \sum_{t=v+1}^{8} \eta_\bullet(\cdot,t,t-v)}, \quad s = 1,\dots,10; \quad v = 2,\dots,7.$$

$$(A.9d)$$

Cancer Modeling with Intermittent Exposures

Daniel Krewski[1] [2] and Duncan J. Murdoch[3]

In this article we consider the application of both the classical Armitage-Doll multi-stage model and the Moolgavkar-Venzon-Knudson two-stage birth-death-mutation model in situations in which carcinogen exposure is not constant over time. In particular, novel representations of the cumulative hazard function are used to describe the relative effectiveness of dosing at different times, and to establish an equivalent constant dose which leads to the same risk as time-dependent dosing. The relative effectiveness function may be used to establish the degree to which the use of a simple time-weighted average dose may underestimate (or overestimate) risk. Both the Armitage-Doll and Moolgavkar-Venzon-Knudson models are applied to bioassay data on $B(a)P$ with variable dosing patterns, using equivalent constant doses to facilitate maximum likelihood estimation of the model parameters.

1. Introduction

Long-term laboratory studies of carcinogenicity are, with some exceptions, conducted with exposure regimens in which dosing is held constant for the duration of the study [7]. Human exposure to carcinogens, on the other hand, may vary markedly, posing the problem of predicting risks associated with variable dosing. For example, changes in human exposure can be due to variation in levels of environmental contamination, occupational changes, or alterations in lifestyle factors such as dietary habits or tobacco consumption.

One approach to cancer risk assessment with variable exposure levels is to base estimates of risk on a time-weighted lifetime average daily dose, calculated by amortizing cumulative lifetime dose on a daily basis. This approach is suggested in the cancer risk assessment guidelines developed by the U.S. Environmental Protection Agency ([18], p.339981) in the absence of data indicating

[1] Environmental Health Directorate, Health Protection Branch, Health and Welfare Canada, Ottawa, Ontario, Canada K1A 0L2

[2] Department of Mathematics and Statistics, Carleton University, Ottawa, Ontario, Canada K1S 5B6

[3] Department of Statistics & Actuarial Science, University of Waterloo, Waterloo, Ontario, Canada N2L 3G1

the existence of dose rate effects. This may require substantial amortization with infrequent exposures, a situation encountered by the National Research Council [14] in setting guidance levels for emergency exposures to chemical carcinogens over periods as short as 24 hours.

In this paper, we examine the consequences of using a time-weighted average dose for cancer risk assessment. This is done within the framework provided by the classical Armitage-Doll multi-stage model of carcinogenesis as well as the more recent Moolgavkar-Venzon-Knudson two-stage birth-death-mutation model. Either of these models can be used to predict cancer risks with intermittent exposures, although the latter model has the advantage of accounting for the effects of tissue growth and cell kinetics in addition to the genetic events involved in carcinogenesis. Using novel representations of the hazard functions for these two models, the relative effectiveness of dosing at different times can be determined.

The Armitage-Doll and Moolgavkar-Venzon-Knudson models may also be applied to experimental data on carcinogenesis in which the dose is not held constant throughout the study period. Using the relative effectiveness functions to define an equivalent constant dose, it is possible to apply maximum likelihood methods developed for constant dosing. This will be illustrated using laboratory data on forestomach tumors in mice exposed to benzo(a)pyrene at different times.

2. Armitage-Doll Multi-Stage Model

The classical Armitage-Doll multi-stage model has been widely applied in the quantitative description of carcinogenesis data [1]. This model is based on the premise that neoplastic transformation of normal tissue requires the occurrence in sequence of $k \geq 1$ fundamental biological events within a single cell. Crump & Howe [6] extended the multi-stage model to the case where the dose $d(u)$ may vary as a function of time u. In this model the cumulative hazard function may be expressed as

$$H(t) = N \frac{t^k}{k!} E \left[\prod_{i=1}^{k} \{a_i + b_i d(U_{i:k})\} \right], \tag{2.1}$$

where N denotes the number of stem cells at risk in the target tissue, $U_{i:k}$ is the ith order statistic in a random sample of size k from the uniform distribution on the interval $[0, t]$ with joint density $k!/t^k$, and the expectation $E(\cdot)$ is taken with respect to this joint density [13].

To examine the effects of time-dependent exposure, consider first the simplest case in which only the transition intensity function for stage $1 \leq r \leq k$ is affected by the agent of interest. In this case the cumulative hazard may be expressed as

$$H(t) = A \frac{t^k}{k!} \{1 + B_r d_r^*(t)\}, \tag{2.2}$$

where $A = N \prod_{i=1}^{k} a_i$, $B_r = b_r/a_r$, and $d_r^*(t) = E[d(U_{r:k})]$ is that level of exposure which, if administered at a constant rate from time 0 to time t, would produce the same cumulative hazard $H(t)$ at time t. Noting that the marginal density of $U_{r:k}$ is given by

$$w(u) = t^{-k} u^{r-1} (t-u)^{k-r} / B(r, k-r+1) \qquad (2.3)$$

$(0 < u < t)$, where $B(\cdot, \cdot)$ denotes the beta function, the equivalent constant dose may be computed as

$$d_r^*(t) = \int_0^t d(u) w(u) du. \qquad (2.4)$$

It follows from (2.2) that at a fixed time t the cumulative hazard $H(t)$ for a multi-stage model with only one stage dose-dependent under time-dependent dosing is of the same general form as that of a one-stage model under constant dosing. It also follows that a constant time-weighted average dose

$$\bar{d}(t) = t^{-1} \int_0^t d(u) du \qquad (2.5)$$

will not generally lead to the same cumulative hazard $H(t)$ as the variable exposure pattern $d(\cdot)$.

To illustrate, consider the case of $k = 6$ with $r = 1$, and suppose that

$$d(u) = \begin{cases} 1, & 0 < u < t/4 \\ 0, & t/4 \le u \le t. \end{cases} \qquad (2.6)$$

In this case, the equivalent constant $d_1^*(t) = 0.83$ exceeds the time-weighted average dose $\bar{d}(t) = 0.25$ by more than three-fold.

In general, the relative effectiveness of dosing at time u may be conveniently represented by the function

$$R(u) = tw(u). \qquad (2.7)$$

This represents a rescaled form of the density function $w(\cdot)$, normalized so that its average value is unity.

The relative importance of exposures occurring between time 0 and time t expressed in terms of $R(\cdot)$ is illustrated in Figure 1 for a k-stage model ($1 \le k \le 6$). Examination of these results demonstrates that early exposures will be more effective than later exposures when an early stage is dose dependent. Conversely, when a late stage is dose-dependent, later exposures will be of greater concern.

Since $d_r^*(t)$ is a weighted average of the function $d(u)$ over $0 < u < t$, the ratio R of this equivalent constant dose to the time weighted average dose $\bar{d}(t)$ will be bounded by the maximum value of the function $R(\cdot)$. Thus, we have

$$R \le \frac{k!(r-1)!(k-r)^{k-r}}{(r-1)!(k-r)!(k-1)^{k-1}} \le k, \qquad (2.8)$$

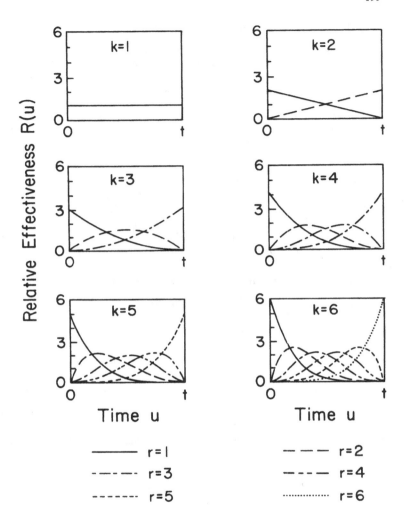

Figure 1

Relative Effectivenss of Dosing at Different Times in a k-Stage Model
$(1 \leq k \leq 6)$ with the rth Stage Dose-Dependent

with the latter inequality noted by Kodell *et al.*[9].

At low doses, the probability of a tumor occurring by time t is approximately equal to the cumulative hazard $H(t)$. Thus, the upper bound R also applies to the degree of underestimation of the excess risk $H(t) - a(t)$, where $a(t) = At^k/k!$ denotes the cumulative hazard at dose zero.

Similar analyses may be conducted when two or more stages are dose-dependent, although the results obtained will generally be too cumbersome to be of practical use. At low doses, however, it can be shown that R remains bounded above by k, regardless of the number of stages affected [13].

These results suggest that the use of a simple time-weighted average dose for risk assessment will generally not understate the risk by a factor of more than k, the number of stages in the model. This has also been demonstrated by Crump & Howe [6] and Kodell *et al.*[9] in specific cases. In the next section we examine the impact of allowing for tissue growth and cell kinetics within the framework provided by the two-stage birth-death-mutation model.

3. Moolgavkar-Venzon-Knudson Two-Stage Model

Moolgavkar [11] describes a more biologically motivated model of carcinogenesis based on the notion that a tumor may be initiated following the occurrence of genetic damage in one or more stem cells in the target tissue as a result of exposure to an initiator. Such initiated cells may then undergo malignant transformation to give rise to a cancerous lesion. The rate of occurrence of such lesions may be increased by exposure to a promotor, which serves to increase the pool of initiated cells through mechanisms resulting in clonal expansion.

Assuming that the target tissue is sufficiently large so that the number of normal cells $N(u)$ present at time u is essentially deterministic, and that the rate of production of cancerous cells is low, the cumulative hazard for this model is of the form

$$H(t) = \int_0^t \int_0^{u_2} \lambda_1(u_1)\lambda_2(u_2)N(u_1)\exp[\tau(u_2) - \tau(u_1)]du_1 du_2 \qquad (3.1)$$

[17], where $\lambda_1(u_1) = a_1 + b_1 d(u_1)$ denotes the rate of occurrence of the first mutation at time u_1, $\lambda_2(u_2) = a_2 + b_2 d(u_2)$ the rate of occurrence of the second mutation at time $u_1 < u_2 < t$, and $\tau(t) = \int_0^t \delta(u)du$, where $\delta(u)$ is the net birth rate of initiated cells at time u. We note that although the expression in (3.1) is widely used, it represents an approximation to the Moolgavkar-Venzon-Knudson-Model which is valid only for low to moderate tumor occurrence rates [3,12]. Although not widely recognized, this also applies to the Armitage-Doll model [12].

As in the previous section, we may write

$$H(t) = E[\lambda_1(U_1)\lambda_2(U_2)]k(t), \qquad (3.2)$$

where $0 < U_1 < U_2 < t$ are random variables with joint density

$$w_{12}(u_1, u_2) = N(u_1) \exp[\tau(u_2) - \tau(u_1)]/k(t), \tag{3.3}$$

and

$$k(t) = \int_0^t \int_0^{u_2} N(u_1) \exp[\tau(u_2) - \tau(u_1)] du_1 du_2. \tag{3.4}$$

To examine the relative effectiveness of exposures at different times consider first the special case in which $N(\cdot) \equiv N$ and $\delta(\cdot) \equiv \delta \neq 0$ are constant. The joint density w_{12} in (3.3) then reduces to

$$w_{12}(u_1, u_2) = \exp[(u_2 - u_1)\delta]/k(t), \tag{3.5}$$

with

$$k(t) = \delta^{-2}(e^{t\delta} - t\delta - 1). \tag{3.6}$$

The marginal distributions of u_1 and u_2 are given by

$$w_1(u_1) = [\delta k(t)]^{-1}[e^{(t-u_1)\delta} - 1] \tag{3.7}$$

and

$$w_2(u_2) = [\delta k(t)]^{-1}[e^{u_2\delta} - 1] \tag{3.8}$$

respectively. As in (2.4), the equivalent constant dose if only stage r is dose dependent is

$$d_r^*(t) = \int_0^t d(u)w_r(u)du. \tag{3.9}$$

The relative effectiveness function when stage $r = 1$ is dose-dependent is shown in Figure 2 for selected values of $t\delta$. (This quantity represents the net birth rate for initiated cells over a time interval of length t.) This example again illustrates that early exposures are more effective than late exposures when an early stage is dose-dependent. Unlike the multi-stage model, however, the relative effectiveness is not bounded by the number of stages in the model, and may be as large as $t\delta$. Chen et al.[4] provide further numerical examples of this behavior.

Similar analyses can be performed if $N(u)$ or $\delta(u)$ vary with time u in a known fashion. In general, these functions will be needed to be determined using special studies. Murdoch & Krewski [13] considered a simple example in which $N(u)$ was proportional to tissue size, using data on human liver weight as an indicator of liver size. This analysis revealed a decrease in relative effectiveness of dosing early on due to the smaller number of cells at risk. In general, however, the number of stem cells at risk may not be directly proportional to the number of cells in the target tissue, since fully differentiated cells may not be good candidates for neoplastic conversion [5].

To examine the effects of variability in $\delta(\cdot)$ over time, suppose that initiated cells proliferate in proportion to the stem cells. Specifically, for $\delta(u) \propto N'(u)/N(u)$ we have

$$\tau(u_2) - \tau(u_1) = C \log\{N(u_2)/N(u_1)\} \tag{3.10}$$

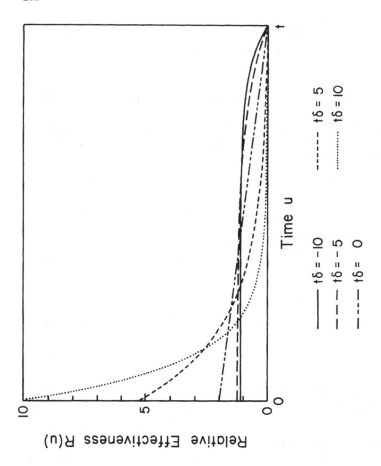

Figure 2

Relative Effectiveness of Dosing at Different Times in the M-V-K Model
with the First Stage Dose-Dependent (N, δ constant)

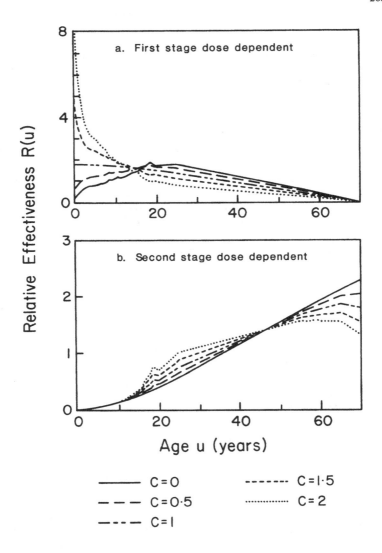

Figure 3

Relative Effectiveness of Dosing at Different Times in the M-V-K Model
with Cellular Proliferation δ Proportional to Human Liver Growth Rate
(The proportionality constant C is defined in (3.10) in the text.)

in (3.3). The relative effectiveness funtion $R(\cdot)$ of dosing when either the first or second stage is dose-dependent is shown for human liver tissue in Figure 3. Since $N(\cdot)$ is increasing early in life, cellular proliferation has a notable impact on relative effectiveness when the first stage is dose-dependent, but has little effect when the second stage depends on dose. (Note that $C = 0$ implies no proliferation of the initiated cells.)

4. Model Fitting

4.1. Maximum Likelihood Estimation

Maximum likelihood methods for fitting dose response models to experimental data on carcinogenesis when the dose remains fixed during the course of the study are well established [8], [10]. To illustrate, suppose that observations (T_i, J_i, d_i) are available for individual $i = 1, \ldots, n$. Here, $J_i = 1$ if the ith individual is observed to have a tumor present at the time T_i of death or sacrifice; otherwise, $J_i = 0$. Exposure to the ith individual is at a constant level d_i.

In what follows, we assume that the intercurrent mortality is independent of the process of tumor induction. Of interest is estimation of the function

$$P(t, d) = 1 - \exp\{-H(t, d)\}, \tag{4.1}$$

which represents the probability of tumor occurrence by time t under constant dosing at level d. If the tumors are not life-threatening, the likelihood is

$$L = \prod_{i=1}^{n} [1 - \exp\{-H(T_i, d_i)\}]^{J_i} [\exp\{-H(T_i, d_i)\}]^{1-J_i}. \tag{4.2}$$

For rapidly lethal tumors, T_i will approximate the time of tumor occurrence, leading to the likelihood

$$L = \prod_{i=1}^{n} \exp\{-H(T_i, d_i)\} [h(T_i, d_i)]^{J_i}, \tag{4.3}$$

where $h = H'$ denotes the hazard function for tumor onset.

4.2. Bioassay of $B(a)P$

Similar likelihoods can be constructed based on the cumulative hazard $H(T, d(\cdot))$ corresponding to the variable dosing pattern $d(\cdot)$, as in Brown & Hoel [2]. A simpler approach is to calculate an equivalent constant dose d^*, and use the likelihood function for constant dosing. Since the equivalent constant dose leads to the same cumulative hazard H as the variable dosing pattern at time T, this will work with the likelihood for incidental tumors which depends only on H, but not with the likelihood for rapidly fatal tumors which depends on the instantaneous hazard h.

This approach may be illustrated using the data in Table 1 on forestomach tumors induced in mice exposed to variable doses of benzo(a)pyrene (B(a)P) originally reported by Neal & Rigdon [15], and recently re-evaluated by Thorslund

Table 1. Forestomach Tumors in Mice Subjected to Variable Exposures of Benzo(a)pyrene[a]

Dose Group	Dose d (mg B(a)P per g food)	Exposure Period t_1-t_2 days	Time at Sacrifice T days	Number of Animals Exposed	Number of Animals with Tumors		
					Observed	Predicted	
						A-D Model	M-V-K Model
1	0	-	300	289	0	0.0	0.0
2	0.001	20-140	140	25	0	0.2	0.2
3	0.01	30-140	140	24	0	1.9	1.4
4	0.02	116-226	226	23	1	3.6	2.7
5	0.03	50-160	160	37	0	8.3	6.3
6	0.04	67-177	177	40	1	11.4	8.8
7	0.05	51-161	163	40	4	13.0	10.4
8	0.05	20-172	172	34	24	18.8	23.8
9	0.10	22-132	132	23	19	13.3	11.1
10	0.25	19-137	137	73	66	67.0	66.6
11	0.25	49-50	155	10	0	0.4	0.4
12	0.25	56-58	162	9	1	0.6	0.8
13	0.25	49-53	155	10	1	1.3	1.6
14	0.25	62-67	168	9	4	1.5	1.7
15	0.25	49-56	155	10	3	2.2	2.5
16	0.25	91-121	198	26	26	16.1	15.5
17	0.10	74-81	182	10	0	1.0	1.1
18	0.10	48-78	156	18	12	5.8	5.6
19	5.00	139-140	252	33	17	17.9	21.9

[a] Adopted from Neal & Rigdon (1967). (In cases where t_1 and t_2 are known only to lie within a certain interval, the midpoint of the interval is shown.)

[16] using direct methods of analysis, assuming the tumors to be incidental. Note that in all cases the exposure patterns are of the form

$$d(u) = \begin{cases} 0, & u < t_1 \\ d, & t_1 \le u \le t_2 \\ 0, & t_2 < u \end{cases} \tag{4.4}$$

where t_1 denotes the age at which exposure started. The actual exposure regimens are depicted in graphical form in Figure 4. (The data reported in Table 1 actually represent an approximation to the actual data since in some cases Neal & Rigdon reported only that t_1 and t_2 lay within a certain interval. The values of t_1 and t_2 given here represent the midpoints of those intervals.)

4.3. Dose Response Models

In his analysis of the Neal & Rigdon data, Thorslund [16] employed a special case of the two-stage birth-death-mutation model in which both the number of stem cells $N(\cdot)$ and the net birth rate of the initiated cell population $\delta(\cdot)$ were held constant over time. Thorslund also assumed that the relative transition rates for the two mutation stages were equal (i.e., $b_1/a_1 = b_2/a_2$). This latter assumption is not necessary and will not be invoked here.

In order to evaluate the effects of allowing for promotion of initiated cells through δ, we will also use the Armitage-Doll two-stage model for comparison purposes. Under this model, Murdoch & Krewski [13] show that the cumulative hazard may be expressed as

$$H(t) = \frac{t^2}{2} \left\{ A + B d_1^*(t) + C d_2^*(t) + \frac{BC}{A} d_{12}^*(t) \right\}, \tag{4.5}$$

where $A = N a_1 a_2$, $B = N b_1 a_2$ and $C = N b_2 a_1$.

For dosing regimens of the form (4.4), the equivalent constant doses required to evaluate (4.5) are given by

$$d_1^*(t) = \int_0^t d(u) w_1(u) du = \begin{cases} 0, & t < t_1 \\ d\left(\frac{t-t_1}{t}\right)^2, & t_1 \le t \le t_2 \\ d\frac{(t_2-t_1)(2t-t_1-t_2)}{t^2}, & t_2 < t, \end{cases} \tag{4.6}$$

since $w_1(u) = 2t^{-2}(t - u)$. Similarly, we find

$$d_2^*(t) = \begin{cases} 0, & t < t_1 \\ d\left(\frac{t^2-t_1^2}{t^2}\right), & t_1 \le t \le t_2 \\ d\left(\frac{t_2^2-t_1^2}{t^2}\right), & t_2 < t \end{cases} \tag{4.7}$$

and

$$d_{12}^*(t) = \begin{cases} 0, & t < t_1 \\ d^2\left(\frac{t-t_1}{t}\right)^2, & t_1 \le t \le t_2 \\ d^2\left(\frac{t_2-t_1}{t}\right)^2, & t_2 < t. \end{cases} \tag{4.8}$$

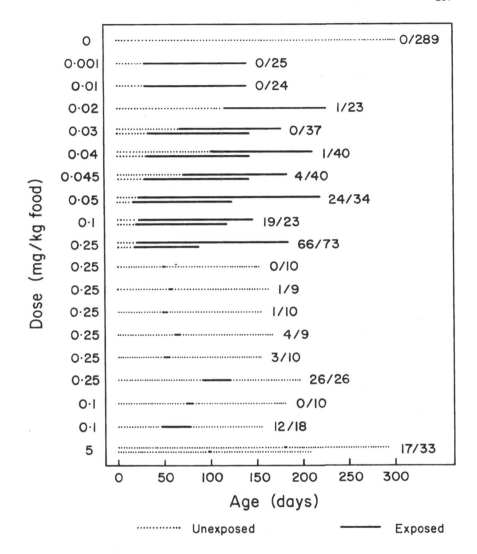

Figure 4

Variable Exposure Regimens used in the Neal-Rigdon Study
(Two exposure regimens corresponding to the earliest and latest possible exposure periods are shown in cases where the actual period is not specified precisely in [15]. Tumor occurrence rates are shown at the end of the exposure period.)

The cumulative hazard in (4.5) may now be used in conjunction with the likelihood for incidental tumors in (4.2) to estimate A, B and C.

The two-stage birth-death-mutation model with N and δ constant can be handled in a similar fashion. Specifically, the cumulative hazard is of the form

$$H(t) = k(t) \left\{ A + Bd_1^*(t) + Cd_2^*(t) + \frac{BC}{A} d_{12}^*(t) \right\}. \qquad (4.9)$$

The equivalent constant doses can be shown to be

$$d_1^*(t) = \begin{cases} 0, & t < t_1 \\ d\frac{k(t-t_1)}{k(t)}, & t_1 \le t \le t_2 \\ d\frac{k(t-t_1)-k(t-t_2)}{k(t)}, & t_2 < t. \end{cases} \qquad (4.10)$$

$$d_2^*(t) = \begin{cases} 0, & t < t_1 \\ d\frac{k(t)-k(t_1)}{k(t)}, & t_1 \le t \le t_2 \\ d\frac{k(t_2)-k(t_1)}{k(t)}, & t_2 < t. \end{cases} \qquad (4.11)$$

$$d_{12}^*(t) = \begin{cases} 0, & t < t_1 \\ d^2 \frac{k(t-t_1)}{k(t)}, & t_1 \le t \le t_2 \\ d^2 \frac{k(t_2-t_1)}{k(t)}, & t_2 < t. \end{cases} \qquad (4.12)$$

Since $k(t) \to t^2/2$ as $\delta \to 0$, these expressions include those given previously for the Armitage-Doll two-stage model as a special case.

Since d_1^*, d_2^* and d_{12}^* in (4.11)-(4.13) involve the unknown parameter δ, through $k(\cdot)$, these equivalent constant doses cannot be evaluated prior to likelihood maximization. Because of this, values of d_1^*, d_2^* and d_{12}^* based on a fixed value for δ were used for maximum likelihood estimation of A, B and C. The likelihood was then maximized by repeating this process over a range of values of δ.

4.4. Results

Both the Armitage-Doll (A-D) model in (4.5) and the Moolgavkar-Venzon-Knudson (M-V-K) model in (4.9) were fit to the Neal-Rigdon data (Table 2, mean times). Comparing the observed and predicted numbers of tumours in Table 1 indicates that neither model provided a good fit to the Neal-Rigdon data. The chi-square goodness of fit statistics were 50.9 on 16 degrees of freedom ($p = 0.00002$) and 44.7 on 15 degrees of freedom ($p = 0.00009$) for the A-D and the M-V-K models respectively due to overestimation of the response in dose groups 6, 7 and 8, and underestimation in groups 10, 17 and 19. Because of this lack-of-fit, no attempt was made to evaluate the standard errors of the estimated model parameters using observed information.

To explore the possibility that our approximation of the start and stop times t_1 and t_2 by the midpoint of the range of times may have introduced substantial errors, we re-fit both models using the maximum values of t_1 and t_2 within the range given for each dose group (Table 2, late times). This reduced the expected number of tumors slightly in the groups in which the response was overestimated, but was not sufficient to obtain an acceptable fit. The use of the minimum values of t_1 and t_2 (Table 2, early times) also failed to provide a better fit to the data. While it is possible that this may be due to the inability of these two models to describe these data, inadequate experimental technique seems more plausible in light of the flexibility of the M-V-K model. This position is supported by the marked difference in response in dose groups 7 and 8 despite almost identical exposure conditions. It is also possible that the approximate forms of both the A-D and M-V-K models used here may be responsible in part for this behavior.

Table 2. Maximum Likelihood Estimates of the Parameters in Two-Stage Models Fitted to the Neal-Rigdon Data

Exposure Times	Parameter	Units	Two-Stage Model A-D	Two-Stage Model M-V-K
Early	A	(days in life)^2	2.1×10^{-14}	2.0×10^{-14}
	B	[mg B(a)P per g food per (days in life)^2]$^{-1}$	0.0011	0.00075
	C	[mg B(a)P per g food per (days in life)^2]$^{-1}$	2.2×10^{-13}	2.4×10^{-13}
	δ	(days in life)$^{-1}$	-	0.0096
Mean	A	(days in life)$^{-2}$	2.4×10^{-14}	3.8×10^{-14}
	B	[mg B(a)P per g food per (days in life)^2]$^{-1}$	0.0014	0.00025
	C	[mg B(a)P per g food per (days in life)^2]$^{-1}$	3.6×10^{-15}	3.7×10^{-14}
	δ	(days in life)$^{-1}$	-	0.029
Late	A	(days in life)$^{-2}$	0	0
	B	[mg B(a)P per g food per (days in life)^2]$^{-1}$	0.0011	0.00075
	C	[mg B(a)P per g food per (days in life)^2]$^{-1}$	0	0
	δ	(days in life)$^{-1}$	-	0.0062

Lack of fit aside, these results do provide some information on the effects of $B(a)P$ within the context of a two-mutation model of carcinogenesis. Although the coefficients b_1 and b_2 reflecting the increased rate of occurrence of these two mutations are not identifiable due to the confounding of a_1 and a_2 in the product $Na_1a_2 = A$, the relative increases in the background mutation rates b_1/a_1 and b_2/b_2 are estimated by B/A and C/A respectively. Based on the analysis of the mean exposure times in Table 2, we estimate $b_1/a_1 \approx 5.8 \times 10^{10}$ and $b_2/a_2 \approx 1.5 \times 10^{-1}$ for the A-D model, indicating that the relative increase in the mutation rate is far greater for the first stage than for the second stage. This also occurs in the analysis of early exposure times. With the late exposure times, the estimated value of $A = 0$ is on the boundary of the parameter space, and is inconsistent with the condition that $a_1a_2 > 0$ implied by the assumption that both first and second stage mutations can occur spontaneously.

The analysis of the M-V-K model again indicates a greater relative effect on the first stage mutation rate than on the second stage using either mean or early exposure times. The positive estimate of δ also suggests that cellular proliferation may also play a role in the induction of forestomach tumors, although the present analysis does not consider this to be related to exposure to $B(a)P$.

The relative effectiveness functions for the fitted A-D and M-V-K two-stage models using the mean exposure times are shown in Figure 5, where the small relative increase in the background mutation rate in the second stage has been ignored. Since only the first stage is considered dose-dependent in both cases, earlier exposures are more effective than later exposures. The allowance for clonal expansion of cells having sustained the first mutation in the M-V-K model results in greater relative effectiveness of dosing early in life in comparison with the A-D model.

The ratios $d_1^*(t)/\bar{d}(t)$ of the equivalent constant dose to the time weighted average dose varied from 0.5 to 1.4 in the A-D model depending on the dose group and varied from 0.07 to 1.1 in the M-V-K model (Table 3). These ratios represent the effectiveness of the actual dosing regimen used in each group in comparison with a constant daily dose obtained by amortizing the total dose administered over the period of observation.

5. Conclusions

Both the Armitage-Doll multi-stage model and the Moolgavkar-Venzon-Knudson two-stage birth-death-mutation model may be used for cancer risk assessment with intermittent exposures. The latter model has the advantage of explicitly allowing for growth of the target tissue and the kinetics of initiated cells. Many chemicals have the ability to induce cellular proliferation, which can markedly increase cancer risk.

In this paper, we have explored the use of these models in predicting cancer risks with time-dependent exposure patterns. Relative effectiveness functions were derived to describe the relative impact on risk of dosing at different times. For both the Armitage-Doll and Moolgavkar-Venzon-Knudson model, early dosing is more effective than later dosing when an early stage is dose-dependent. Conversely, later exposures are of greater concern with a late stage carcinogen.

To examine the impact of changes in the number of normal cells over time, it is necessary to specify the growth pattern of target tissue. Using growth curves

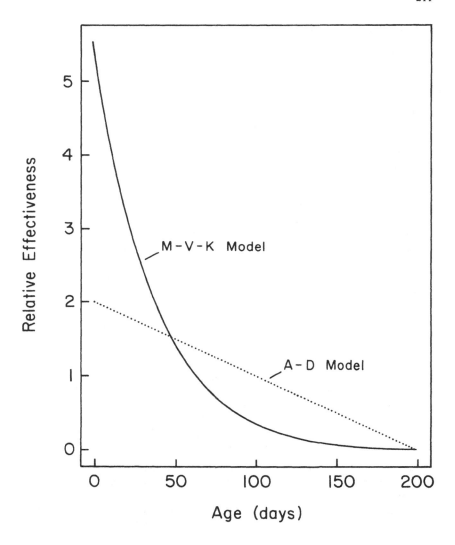

Figure 5

Relative Effectiveness of Dosing in the A-D and M-V-K Two-Stage Models
Fitted to the Neal-Rigdon Data (for tumor occurrence by 200 days)

Table 3. Ratio of Equivalent Constant and Time-Weighted Average Doses
Based on Two-Stage Models Fitted to the Neal-Rigdon Data

Exposure Period	Age at Sacrifice	Ratio of Doses $d_1^*(T)/d(T)$	
$t_1 - t_2$ days	T days	A-D Model	M-V-K Model
30-140	140	.79	.50
30-140	140	.79	.50
116-226	226	.49	.07
50-160	160	.69	.32
67-177	177	.62	.22
51-161	163	.70	.32
20-172	172	.88	.64
22-132	132	.83	.60
19-137	137	.86	.65
49-50	155	1.36	1.12
56-58	162	1.30	.94
49-53	155	1.34	1.08
62-67	168	1.23	.79
49-56	155	1.32	1.03
91-121	198	.93	.29
74-81	182	1.15	.59
48-78	156	1.19	.79
139-140	252	.89	.15

for human liver tissue as an illustration, it was seen that when the net growth
rate of initiated cells is constant over time, early dosing remains most effective
when the first stage is dose dependent with later dosing more effective when the
second stage is dose dependent. In comparison with the case where the target
tissue was considered to be of constant size, the relative effectiveness of early
dosing is reduced when the first stage is dose dependent because of the reduced
number of liver cells available for transformation early in life.

The impact of the assumption that the net birth rate of initiated cells is
constant over time was also examined in this example by allowing the initiated
cells to proliferate at a rate proportional to that of normal liver cells. In this
case, the effects attributable to tissue growth noted above are reduced, with the
relative effectiveness of dosing depending on the relative rates of growth of the
normal and initiated cell populations.

It was also shown that the risk predicted on the basis of a simple time-
weighted average level of exposure will generally differ from the actual risk
corresponding to the particular time-dependent exposure pattern experienced.
Under the multi-stage model, this difference will be limited to a factor of k,
the number of stages in the model. Under the two-stage birth-death-mutation
model, however, the use of a constant time-weighted average dose can seriously
underestimate risk in the presence of rapid proliferation of the initiated cells.

The relative effective functions developed here can be used to define equivalent constant doses which lead to the same risk as variable dosing. The use of

equivalent constant doses facilitates model fitting in that maximum likelihood methods developed for constant dosing can be exploited. This was illustrated using experimental data on benzo(a)pyrene involving different exposure periods. Although neither the A-D nor the M-V-K model provided good fits to these data, this may be due to experimental problems rather than the inapplicability of these two models. This lack-of-fit notwithstanding, the relative increase in the first stage mutation rate appeared to be far greater than in the second stage, leading to greater relative effectiveness of early as compared to late exposures. Allowance for proliferation of initiated cells (unrelated to $B(a)P$) in the M-V-K two-stage model led to higher relative effectiveness of dosing early in life in comparison with the A-D two-stage model.

6. Acknowledgements

This research was supported in part by grant nos. A8664 and OGPIN 0/4 from the National Sciences and Engineering Research Council of Canada to D. Krewski and D. Murdoch respectively.

7. References

1. Armitage, P. (1985). Multistage models of carcinogenesis *Environmental Health perspectives 63*, 195-201.

2. Brown, K.G. & Hoel, D.G. (1986). Statistical modeling of animal bioassay data with variable dosing regimens: example - vinyl chloride. *Risk Analysis 6*, 155-166.

3. Chen, C.W. & Moini, A. (1989). On cancer dose response models with clonal expansion. In: *Scientific Issues in Quantitative Cancer Risk Assessment* (S.H. Moolgavkar, ed.). Birkhauser Boston, New York. In Press.

4. Chen, J.J., Kodell, R.L. & Gaylor, D. (1988). Using the biological two-stage model to assess risk from short-term exposures. *Risk Analysis 6*, 223-230.

5. Clifton, K-H. (1989) The clonogenic cells of the rat mammary and thyroid glands: their biology, frequency of initiation, and promotion/progression to cancer. In: *Scientific Issues in Quantitative Cancer Risk Assessment* (S.H. Moolgavkar, ed.). Birkhauser Boston, New York. In Press.

6. Crump, K.S. & Howe, R.B. (1984). The multistage model with a time-dependent dose pattern: applications to carcinogenic risk assessment. *Risk Analysis 4*, 163-176.

7. Gart, J.J., Krewski, D., Lee, P.N., Tarone, R.L. & Wahrendorf, J. (1986). *Statistical Methods in Cancer Research, Vol. II, The Design and Analysis of Long-Term Animal Experiments.* IARC Scientific Publications No.79, International Agency for Research on Cancer, Lyon.

8. Kalbfleisch, J.D.; Krewski, D.R. and Van Ryzin, J. (1983). Dose-response models for time-to-response toxicity data (with discussion by V.T. Farewell and J.F. Lawless). *Canadian Journal of Statistics 11*, 25-49.

9. Kodell, R.L., Gaylor, D.W. & Chen, J.J. (1987). Using average lifetime dose rate for intermittent exposures to carcinogens. *Risk Analysis 7*, 339-345.

10. Krewski, D., Murdoch, D. & Dewanji, A. (1986). Statistical modeling and extrapolation of carcinogenesis data. In: *Modern Statistical Methods in Chronic Disease Epidemiology* (S.H. Moolgavkar & R.L. Prentice, eds.). Wiley-Interscience, New York, pp. 259-282.

11. Moolgavkar, S.H. (1986). Carcinogenesis modeling: from molecular biology to epidemiology. *Annual Review of Public Health 7*, 151-169.

12. Moolgavkar, S. & Dewanji, A. (1988). Biologically based models for cancer risk assessment: a cautionary note. *Risk Analysis 8*, 5-6.

13. Murdoch, D.J. & Krewski, D. (1988). Carcinogenic risk assessment with time-dependent exposure patterns. *Risk Analysis 8*, 521- 530.

14. National Research Council (1988). Criteria and Methods for Preparing Emergency Exposure Guidance Level (EEGL), Short-Term Public Emergency Guidance (SPEGL), and Continuous Exposure Guidance Level (CEGL) Documents. National Academy Press, Washington, D.C.

15. Neal, J. & Rigdon, R.H. (1967). Gastric tumors in mice fed benzo[a]pyrene: a quantitative study. *Texas Reports in Biology and Medicine 24*, 553-557.

16. Thorslund, T. (1988). Comparative Potency Approach for Estimating the Cancer Risk Associated with Exposure to Mixtures of Polycyclic Aromatic Hydrocarbons. Clement Associates, Washington, D.C.

17. Thorslund, T.W., Brown, C.C. & Charnley, G. (1987). Biologically motivated cancer risk models. *Risk Analysis 7*, 109-119.

18. U.S. Environmental Protection Agency (1986). Guidelines for carcinogen risk assessment. *Federal Register 51*, 33992-34003.

POSSIBLE ROLE OF SELECTIVE GAP JUNCTIONAL INTERCELLULAR COMMUNICATION
IN MULTISTAGE CARCINOGENESIS

Hiroshi Yamasaki and D. James Fitzgerald

Abstract

Carcinogenesis modelling must incorporate current knowledge of the interaction between "initiated" preneoplastic cells and surrounding normal cells. In this context, an important mode of cell:cell interaction is that of gap-junctional intercellular communication (GJIC); in normal tissue, GJIC plays a key role in maintaining homeostasis, so it is therefore not surprising to find GJIC disturbance in cancer cells. Furthermore, since many tumour promoters can inhibit GJIC, it is possible that such inhibition is critical to the process of carcinogenesis. Advancing our understanding of GJIC disruption and how to reverse it should be useful in application to anti-carcinogenesis and tumour suppression.

1. Introduction

During multistage carcinogenesis, a cell undergoes a series of genetic as well as epigenetic changes and deviates from the normal controls of proliferation and differentiation. Normal cell growth is controlled by various forms of cell-cell interaction and communication at the cellular level [37], while at the gene level, certain cellular oncogenes are considered to play an important role in such growth control [10]. Thus it is reasonable to assume that the functions of cellular oncogenes and of cell-cell interactions are altered during multistage carcinogenesis. The mathematical model proposed by Moolgavkar and Knudson [38] is based on the biology of multistage carcinogenesis and it emphasizes the importance of the balance of cell proliferation and differentiation. Therefore, the basic mechanistic information presented here should also help in refining such a model.

Our working hypothesis on the mechanism of multistage carcinogenesis is as follows: an initiating agent activates cellular oncogene(s) or other genes; however, mutated cells remain dormant since there is intact gap junctional intercellular communication

(GJIC) with surrounding normal cells and these initiated cells are prevented from clonal expansion; when tumour promoting agents block GJIC, the altered gene structure can be expressed since growth control is no longer exerted via cell communication from surrounding normal cells [60]. A series of experiments conducted by Balmain's group [6, 43] and others [11] with the mouse two-stage carcinogenesis model indeed provide support for the idea that oncogene activation by a carcinogen plays an important role in initiation of carcinogenesis. For example, DNA from mouse skin papillomas and carcinomas produced by an initiation-promotion protocol had an A to T transversion at the 61st codon of the H-ras gene only when 7,12-dimethylbenz(a)-anthracene (DMBA) or dibenz(c,h)acnidine were used as initiators. When benzo(a)pyrene or N-methyl-N'-nitro-N-nitrosoguanidine were used, there was no such mutation at the 61st codon of H-ras, although many tumours contained DNA that transformed NIH 3T3 cells. The fact that an A to T transversion is even found in skin papillomas induced by painting DMBA/TPA suggests that this mutation is an early event. In addition, the same specific mutation was observed in tumours which were produced with DMBA plus other tumour promoters, i.e., chrysarobin and benzoyl peroxide, suggesting that the oncogene activation is specific for initiating agents and not for promoting agents [43, 11].

We have seen similar results using a different exposure protocol in which a single administration of DMBA to mice in utero, at doses which do not induce carcinomas in the progeny, is followed by repeated postnatal applications of 12-O-tetradecanoylphorbol-13-acetate (TPA) to the skin of the progeny. In our experiments, only half the papillomas tested showed the presence of the H-ras 61st codon mutation, whereas all carcinomas possessed this mutation [63]. It may be that papillomas without ras mutations are those induced by TPA and destined to regress, while papillomas with the ras mutation are programmed for progression to carcinomas. But it is also feasible that ras activation can occur spontaneously at later stages and contribute to the progression of the neoplasm toward malignancy.

While cellular oncogene activation can also occur during late stages of carcinogenesis [19, 24], experimental results strongly suggest that "initiated" cells which harbour an activated oncogene remain dormant. In the case of mouse skin experiments, only in the presence of a tumour promoting stimulus do such dormant cells clonally expand to form tumours. Here we will discuss the possible role of blocked cell-cell communication, especially of gap junctional intercellular communication (GJIC), in the clonal expansion of "initiated" cells, i.e., tumour promotion. This mode of cell-cell communication involves exchange, via the membrane-spanning gap junction channels, of ions and low molecular weight (<1,000) molecules directly from one cell interior to the next, and is important in maintaining homeostasis and tissue cooperativity [37].

2. Intercellular communication and tumour promotion

The idea that disturbed GJIC is involved in the process of carcinogenesis is not new. However, there was no evidence for this involvement until it was discovered that tumour promoting phorbol esters (e.g., TPA) are potent inhibitors of GJIC [39, 66]. The inhibition of GJIC between cultured cells was first discovered using metabolic cooperation as an assay to measure GJIC. This finding was soon confirmed by various other investigators and was extended by using different assays of GJIC (Table 1).

It is important to emphasize that not only phorbol esters, but also other types of tumour promoting agents were shown to inhibit GJIC. As shown in Table 1, these tumour promoting stimuli include physical stimuli such as partial hepatectomy and skin wounding. Recently, the systemic administration of liver tumour promoting agents, such as phenobarbital, was shown to reduce the level of gap junction protein mRNA and the number of gap junctions in rat liver [34, 51], and the painting of mouse skin with tumour promoting agents reduced the number of gap junctions in epidermal cells [26]. Recently, however, Kam and Pitts [27] reported that one application of (TPA) did not inhibit GJIC of epidermal cells in intact mouse skin. These results are inconsistent with those obtained from cultured mouse epidermal cells [40, and unpublished observations]. It may be though that multiple TPA applications are needed to reveal the effect in vivo.

Table 1. Inhibition of gap-junctional intercellular communication by tumour promoting stimuli (modified from ref. 57)

Method of communication measurement	Promoting stimulus
Metabolic cooperation	
3H-uridine metabolites transfer	Phorbol esters, chlordane
HGPRT$^+$/HGPRT$^-$(a)	Phorbol esters, PCBs, and many other tumour promoting agents
ASS$^-$/ASL$^-$(b)	Phorbol esters, DDT
AK$^+$/AK$^-$ (c)	Phorbol esters
Electrical coupling	
	Phorbol esters
	Skin wounding
Dye transfer	
Microinjection	Phorbol esters, cigarette smoke condensate, diacylglycerol and other tumour promoting agents
	Partial hepatectomy
Photobleaching	TPA, dieldrin, PBB
Scrape loading	TPA, dieldrin and other tumour promoting agents
Gap junction structure analysis	
Electron microscope	Phorbol esters, mezerein, cigarette smoke condensate
	Phenobarbital, DDT
Gel electrophoresis analysis	Phorbol esters
Analysis with gap junction antibody	Partial hepatectomy
Gap junction gene expression	
Connexin 32mRNA level	Phenobarbital
	Partial hepatectomy

(a) HGPRT, hypoxanthine guanine phosphoribosyltransferase

(b) ASS$^-$, argininosuccinate synthetase-deficient;
    ASL$^-$, argininosuccinate lyase-deficient

(c) AK, adenosine kinase

The inhibition of GJIC by various types of tumour promoting stimuli is the most convincing and first direct evidence to support a role for inhibited communication in the process of carcinogenesis. Further related evidence is that anti-tumour-promoting agents such as retinoic acid, cAMP and glucocorticoids, can antagonize the effect of TPA on the inhibition of GJIC [59].

Mechanisms of GJIC inhibition by tumour promoting agents appear to involve both transcriptional and post-translational control of gap junctions. For example, TPA inhibition of GJIC does not require RNA or protein synthesis, suggesting post-translational mechanisms [61]. Protein kinase C phosphorylates gap junction proteins [52], but it is not clear whether such phosphorylation is the mechanism for decreased GJIC capacity. On the other hand, phenobarbital decreases the level of gap junction mRNA, suggesting a transcriptional control [34].

3. Role of inhibited intercellular communication in cell transformation

The ability of phorbol esters to act as tumour promoters in vitro in two-stage cell transformation assays correlates well with their ability to inhibit GJIC. For example, phorbol-12,13-didecanoate both promotes BALB/c 3T3 cell transformation and inhibits GJIC to a greater degree than TPA [18].

Use of BALB/c 3T3 cell variants has provided another line of evidence for the involvement of blocked GJIC in cell transformation. When a variant (clone A31-1-13) that is sensitive to induction of transformation by chemical carcinogens or UV irradiation was compared to a transformation-resistant variant (clone A31-1-8), no difference was seen in their ability to metabolize carcinogens, to bind these metabolites to DNA, to repair DNA damage, or in their susceptibility to induction of mutation [25, 30]. However, the two variants showed a drastic difference in intercellular communication at cell confluence whereby transformation-sensitive but not transformation-resistant cells became virtually communication-incompetent [62]. We interpret these results as indicating that transformation-sensitive cells have an intrinsic ability to express a TPA-related effect, namely, inhibition of GJIC at confluence, which may be causal in the transformation of this cell line.

There is also a good correlation between inhibition of cell transformation and enhancement of GJIC for chemicals such as cyclic AMP, retinoic acid and glucocorticoids which have been reported to inhibit tumour promotion in mouse skin [48]. We have shown that when these compounds are added to BALB/c 3T3 cells during phorbol ester promotion of transformation, there is a significant decrease in the yield of transformed foci [65]. Previously, we had demonstrated that these compounds can antagonize phorbol ester-mediated inhibition of GJIC between cultured cells [59]. These results are consistent with the idea that tumour promoted-induced inhibition of GJIC is a mechanism for enhancement of cell transformation, and that anti-promoters operate by antagonizing this inhibition of communication.

We have described here only supportive evidence for the involvement of inhibition of GJIC in the enhancement of cell transformation; however, there are results that do not support this working hypothesis. For example, we have shown recently that transforming growth factor β (TGF-β) can significantly enhance BALB/c 3T3 cell transformation in the two-stage protocol; indeed, the effect of TGF-β was greater than that of phorbol esters. However, TGF-β did not inhibit GJIC during the entire period of cell transformation [20]. Boreiko's group have shown that 2,3,7,8-tetrachlorodibenzo-p-dioxin (TCDD) can promote transformation of C3H10T1/2 cells [1], but no inhibition of GJIC was observed [12]. These results suggest that there are multiple mechanisms for enhancement of cell transformation by tumour-promoting agents. It is also possible, however, that there can exist a local inhibition of GJIC around initiated cells, and since there would be only a few initiated cells in any transformation system, we would be unable to detect any significant inhibition of intercellular communication in our assay using TGF-β or TCDD.

A corollary of our hypothesis is that, when cells are transformed, they should not communicate with surrounding normal cells, for if they did communicate, there should be some control exerted by normal surrounding cells via gap junctions and therefore transformed phenotypes would not be maintained. The absence of GJIC between transformed and normal cells has been clearly shown in the BALB/c 3T3 cell transformation system [17, 64]. When transformed foci were obtained by different carcinogens, viz., chemicals, UV or oncogenes, there was a clear selective intercellular communication inhibition between transformed and normal cells; there was GJIC among transformed cells themselves or among normal cells themselves, but not between the two cell types. Similar results were obtained with normal and transformed rat liver epithelial cells [36].

Such a selective inhibition of GJIC between transformed and normal cells was, however, not observed in other transformed cells. For example, when transformed mouse epidermal cells were cocultured with normal cells, there was heterologous intercellular communication [28]. However, when the intrinsic GJIC of the cell lines was examined, their communication capacity was seen to decrease as the cells became more malignant [28], and thus we consider that there may be a decrease of intercellular communication with surrounding normal cells in vivo. Therefore, we propose that there are two ways for transformed cells to attain selective intercellular communication inhibition with normal surrounding cells: 1) by a decrease of intrinsic GJIC capacity such as in the transformed mouse epidermal cells, or 2) by attaining a communication selectivity whereby they maintain intrinsic GJIC capacity but lose the capacity for GJIC with normal cells, such as in BALB/c 3T3 cells and rat liver epithelial cells.

4. Oncogene expression and gap junctional intercellular communication

In our working hypothesis described above, "initiated" cells with altered oncogenes are to be influenced by disturbed GJIC during the tumour promotion phase. To investigate the possible role of GJIC in the regulation of oncogene expression, we have conducted a series of experiments in collaboration with Drs Bignami and Tato. We first prepared various v-onc-containing NIH 3T3 cell clones, the oncogenes being ras, src, myc, fos, polyoma large T and polyoma middle T.

Transfection of these oncogenes into NIH 3T3 cells did not significantly change the homologous GJIC capacity of these cells. However, when they were mixed with normal BALB/c 3T3 cells, some clones showed a heterologous communication capacity while others did not. Cells containing myc, fos or polyoma large T did communicate with surrounding normal cells, and, moreover, these clones did not form distinct foci on the monolayer of normal cells (unless TPA was present). Conversely, ras- or src-containing cells did form distinct foci over monolayers of normal cells and did not show heterologous communication with surrounding normal cells. When clones containing polyoma middle T genes were analyzed, the transfected cells did form a certain number of distinct foci on normal cell monolayers and only low heterologous communication was observed [9, and unpublished observations]. These results clearly suggest that GJIC between normal cells and oncogene-containing cells can influence oncogene-mediated expression of transformed phenotypes.

The above results with oncogene-containing cells further indicate an interesting pattern. Namely, the absence of heterologous GJIC and concomitant foci formation are associated with cytoplasmic/membrane oncogenes (ras, src, polyoma middle T), while the presence of heterologous GJIC and subsequent lack of foci formation are associated with nuclear oncogenes (myc, fos, polyoma large T). More studies are required to determine whether this relationship represents a general phenomenon.

In addition to these studies showing that GJIC can modulate oncogene expression, several reports demonstrate that certain oncogenes can modulate GJIC in various cell types. A summary of this work is presented in Table 2. Further work should reveal molecular mechanisms by which the signals of cell-cell interaction (membrane phenomenon) can be transmitted to the nucleus to control the expression of genes which are critically involved in carcinogenesis.

Table 2. Effect of oncogenes on homologous and heterologous gap
junctional intercellular communication

Oncogene	Cells	Homol. comm.[1]	Hetergl. comm.[2]	Refs
v-src	NRK	↓	NT	Atkinson et al. 1981 [2]
	NIH 3T3	↓	NT	Chang et al., 1985 [13]
	Quail and chick embryo fibrobl.	↓	NT	Azarnia & Loewenstein, 1984 [3]
	NIH 3T3	→	–	Bignami et al.,1988[9]
c-src	NIH 3T3	↓	NT	Azarnia et al.,1988[5]
v-ras	NIH 3T3	→	–	Bignami et al.,1988[9]
EJ-ras[H]	BALB/c 3T3	→	–	Yamasaki et al., 1987 [64]
	Rat liver epith. cell line IAR20	→	+	Enomoto et al., 1987 [16]
	Rat liver epith. cell line	↓	NT	Vanhamme et al., 1989 [54]
v-myc	NIH 3T3	→ or ↓	+	Bignami et al.,1988[9]
v-fos	NIH 3T3	→	+	Unpublished results
v-mos	C3H10T1/2	→ or ↑	NT	Mehta et al., 1989 [33]
PyMT	Rat F cells	↓	NT	Azarnia & Loewenstein 1987 [4]
	NIH 3T3	→	–	Unpublished results
PyLT	NIH 3T3	→	+	Unpublished results
SV40T	Human hepatocytes	↓	NT	Unpublished results
	Human keratino- cytes	↓	NT	Unpublished results
	Human fibroblasts	↓	NT	Rosen et al., 1988 [44]

[1]Homologous communication is the communication among
oncogene-containing cells and their communication capacity was
compared with that of normal counterparts.

[2]↓, decreased; →, no change; ↑, enhanced.

Heterologous communication is the presence (+) or absence (–) of
communication between oncogene-containing cells and normal cells
measured in coculture.
NT, not tested.

5. Intercellular communication and tumour suppression

Our idea on the selective lack of communication of transformed cells with surrounding normal cells as a key element in the maintenance of transformed phenotypes is based on the assumption that normal phenotypes are dominant over transformed ones. This in turn suggests that, if we establish intercellular communication between transformed and surrounding normal cells, then the transformed cells may disappear, thus accomplishing tumour suppression.

There are several examples in which transformed cells did not grow when cocultured with an excess number of normal counterparts. The first description of this type of study came from Stoker and his colleagues [50]. When a small number of polyoma virus-transformed BHK21 cells was cocultured with non-transformed mouse fibroblasts, there was suppression of growth of the transformed cells; direct cell contact was necessary to attain the suppression. Subsequent studies have provided evidence that there was indeed passage of molecules (metabolic cooperation) from surrounding normal cells to transformed cells [49]. Thus, these results suggest that a direct cell contact-mediated transfer of growth control factors from normal cells to transformed cells may suppress the phenotype of transformed cells. Such a suppression was also observed with polyoma virus-transformed bovine fibroblasts cocultured with normal fibroblasts [41].

Sivak and Van Duuren [47] have reported that the outgrowth of SV40-transformed Swiss 3T3 cells was suppressed by coculture with a vast number of normal cells. The addition of croton oil, however, rescued the growth of SV40-transformed cells in coculture. Since croton oil contains TPA and other phorbol esters which are potent inhibitors of GJIC, it is possible that GJIC between transformed and non-transformed cells was the cause of the suppression of this transformed phenotype.

A series of experiments performed by Bertram and his colleagues [7, 8, 31] also suggested that normal cells can suppress transformed phenotypes through cell contact. When transformed C3H10T1/2 cells were cocultured with normal counterparts in the presence of a cAMP phosphodiesterase inhibitor, there was suppression of the transformed phenotype. The suppression was dependent upon close contact of the two cell types and did not appear to involve extracellular factors. It was suggested that an increase of endogenous cAMP levels enhanced the ability of gap junctions to transfer growth-inhibitory regulatory molecules.

Recently, Herschman and Brankow [23] have shown that C3H10T1/2 cells cloned from transformed foci induced by UV radiation and TPA appeared transformed in homologous culture. However, when cocultured with normal cells, they were unable to form foci. In the presence of TPA though, the transformed cells formed foci in these mixed cultures while retinoic acid could block this action of TPA. These results again show that normal cells surrounding transformed cells can suppress the transformed phenotype.

Suppression of transformed cells by close contact with normal cells was also observed after in vivo inoculation of mixed cell populations. Terzaghi-Howe [53] inoculated cell mixtures containing normal and neoplastic rat tracheal epithelial cells into the lumen of denuded rat tracheas, then transplanted these tracheas subcutaneously into syngeneic hosts; an intact tracheal mucosa was regenerated from the inoculated cells within 2-3 weeks after the transplantation. In this system, no tumours developed over a 16-week period provided a minimum number of normal cells was employed in the tracheal repopulation. In contrast, in denuded tracheas containing neoplastic cells alone or in partially denuded tracheas containing comparable numbers of neoplastic cells covering a contiguous area of submucosa, tumours developed 2-5 weeks after cell inoculation and transplantation. These results suggest that normal tracheal epithelial cells could suppress the expression of the neoplastic phenotype in neoplastic populations only when there existed a certain degree of heterologous cell-cell contact. This was provided in populations that were dispersed but not in those that were contiguous.

In a similar study, Dotto et al. [14] demonstrated the suppression of tumourigenicity of transformed mouse epidermal cells by normal fibroblasts. Mouse primary keratinocytes were malignantly transformed

by infection with Harvey sarcoma virus; these cells form carcinomas
when skin-grafted onto syngeneic animals. However, expression of their
malignant phenotype was inhibited when these cells were grafted
together with normal dermal fibroblasts. The suppression was also seen
with mitomycin C-treated dermal fibroblasts but not with untransformed
keratinocytes or 3T3 fibroblasts. It is conceivable that specific
interaction with dermal fibroblasts was responsible for the
suppression of these H-ras-transformed epidermal cells. Earlier
studies of Hennings et al., however, suggested that the interaction
with normal keratinocytes was important for growth suppression of a
papilloma cell line [22].

If selective intercellular communication with non-transformed
cells is critical for the maintenance of transformed phenotypes, the
recovery of communication between transformed cells and normal cells
should cause disappearance of transformed phenotypes. We have recently
performed such an experiment [65]. Cultures containing transformed
BALB/c 3T3 foci on a normal cell monolayer were exposed to cAMP,
retinoic acid or glucocorticoids. These chemicals are known to
upmodulate GJIC but also cause transformed cells to communicate with
normal cells; subsequently, we observed the disappearance of the
transformed foci. Retinoids were recently reported to prevent skin
cancer in xeroderma pigmentosum patients [29]. It is also interesting
to note that retinoic acid causes a proliferation of gap junctions
when applied to a rabbit keratoacanthoma and human basal cell
carcinoma [42, 15]. These results are consistent with the idea that
GJIC plays an important role in the process of carcinogenesis and in
the maintenance of transformed phenotypes.

In considering cancer in human populations, the foregoing
discussion of tumour suppression may indicate one approach for future
studies on dietary or synthetic chemoprevention and chemotherapy. By
understanding which dietary factors may contribute to a more robust
complement of gap junctions (i.e., more gap junctions, or gap
junctions with a decreased sensitivity to xenobiotic disturbance), we
may be able to "naturally" suppress clonal expansion of initiated
cells even in the presence of environmental or occupational tumour
promoters. As regards chemotherapy, dietary factors or synthetic drugs

that enhance GJIC could be efficacious in inducing regression of existing neoplasms.

6. Conclusion and discussion

Evidence for the involvement of selective GJIC in carcinogenesis is rather strong as reviewed above. There is also evidence that restoration of GJIC plays a role in tumour suppression.

However, although GJIC is a good candidate as a major control mechanism of tumour growth, there are other possible forms of interaction among cells which may be important as well. For example, we have recently shown that the gene expression of rat hepatocytes can be regulated by direct contact with cocultured rat biliary epithelial cells, but in this case there was no heterologous GJIC [35]. It is likely that cell-cell interaction at their membrane surface is enough for cells to regulate gene expression; heterologous cell:cell recognition molecules may include certain cell adhesion molecule (CAM) proteins, since it has recently been shown that these proteins (e.g., A-CAM and L-CAM) recognize each other to form functional adherent junctions [55]. Moreover, isolated surface glycoproteins from quiescent cells inhibit the growth of endothelial cells [21] or of human fibroblasts [56]. These results suggest that cell-cell recognition molecules may be important in controlling gene expression and therefore tumour growth. It is also likely that gap junction formation between adjacent cells itself is controlled by such molecules; this has recently indeed been shown. When an expression vector of a cell adhesion molecule (L-CAM) gene was transfected into cells which were deficient in GJIC, they started to communicate through gap junctions [32]. Other membrane components are likely to be involved in the regulation of gap junction formation and function.

Ultimately, it is important to identify growth control factors which pass through gap junctions to control tumour cell growth. Evidence for the existence of tumour suppressing factors in cytoplasm has been recently provided by Shay and his colleagues using 'recons' of transformed cell nuclei and normal cell cytoplasm [46]. Karyoplasts

(nuclei surrounded by a thin layer of cytoplasm within a plasma
membrane envelope) from tumourigenic NIH 3T3 cells were fused to
non-tumourigenic NIH 3T3 cytoplasts to produce 'recons'. Ten clones of
such recons were isolated and all of them were non-tumourigenic. These
results confirm an earlier finding with cybrids between tumour cell
and normal cell cytoplast. While not all other studies support this
finding [45], these results clearly indicate the presence of
cytoplasmic factors in non-tumourigenic cells that can suppress
tumourigenicity. Whether such factors are the same as those putative
regulatory factors that go through gap junctions is presently unknown.
Identification of gap junction-permeant growth regulatory factors is a
difficult task, since their amount in a given cell is probably
minuscule and since there is no simple method to assay their function.
Candidate molecules include cAMP, calcium, phosphatidyl inositol cycle
members and chalones. Chalones, negative growth factors, are
considered to be present in cytoplasm and their molecular weight is
within the exclusion limit of gap junctions. The identification of
such growth control factors may eventually reveal a new family of
tumour suppressor genes.

GJIC is probably an important factor to be considered for
mathematical modelling and risk estimation in multistage
carcinogenesis, which are the main themes of the meeting. GJIC plays
an important role in determining the rate of growth and expansion of
potential tumour cells, these being key parameters in such modelling
[38]. In terms of risk assessment, we may be able to use GJIC assay as
an endpoint for detecting tumour promoting activity of environmental
chemicals [58].

ACKNOWLEDGEMENTS

We thank Ms Chantal Fuchez for her skillful secretarial work. The
work from our laboratory has been partly supported by research grants
from EEC (EV4V1/00182) and from NCI, USA (RO1 CA40534).

REFERENCES

1. Abernethy DJ, Greenlee WF, Huband JC, Boreiko CH (1985)
    2,3,7,8-tetra- chlorodibenzo-p-dioxin (TCDD) promotes the
    transformation of C3H10T1/2 cells. Carcinogenesis, 6, 651-653

2. Atkinson MM, Menko AS, Johnson RG, Sheppard JR, Sheridan JD (1981)
    Rapid and reversible reduction of junctional permeability in cells
    infected with a temperature-sensitive mutant of avian sarcoma
    virus. J. Cell Biol., 91, 573-578

3. Azarnia R, Loewenstein WR (1984) Intercellular communication and
    the control of growth: X. Alteration of junctional permeability by
    the src gene. A study with temperature-sensitive mutant Rous
    sarcoma virus 40. J. Membrane Biol., 82, 191-205

4. Azarnia R, Loewenstein WR (1987) Polyomavirus middle T antigen
    downregulates junctional cell-to-cell communication. Mol. Cell.
    Biol., 7, 946-950

5. Azarnia R, Reddy S, Kmiecik TE, Shalloway D, Loewenstein WR (1988)
    The cellular src gene product regulates junctional cell-to-cell
    communication. Science, 239, 398-401

6. Balmain A, Pragnell IB (1983) Mouse skin carcinomas induced
    in-vivo by chemical carcinogens have a transforming Harvey-ras
    oncogene. Nature, 303, 72-74

7. Bertram JS, Bertram BB, Janik P (1982) Inhibition of neoplastic
    cell growth by quiescent cells is mediated by serum concentration
    and cAMP phosphodiesterase inhibitors. J. Cell. Biochem., 18,
    515-538

8. Bertram JS, Faletto MB (1985) Requirements for and kinetics of
    growth arrest of neoplastic cells by confluent 10T1/2 fibroblasts
    induced by a specific inhibitor of cyclic adenosine
    3':5'-phosphodiesterase. Cancer Res., 45, 1946-1952

9. Bignami M, Rosa S, Falcone G, Tato F, Katoh F, Yamasaki H (1988)
    Specific viral oncogenes cause differential effects on
    cell-to-cell communication, relevant to the suppression of the
    transformed phenotype by normal cells. Mol. Carcinogenesis, 1,
    67-75

10. Bishop JM (1987) The molecular genetics of cancer. Science, 235,
    305-311

11. Bizub D, Wood AW, Skalka AM (1986) Mutagenesis of the Ha-ras
    oncogene in mouse skin tumors induced by polycyclic aromatic
    hydrocarbons. Proc. Natl. Acad. Sci. USA, 83, 6048-6052

12. Boreiko CJ, Abernethy DJ, Sanchez JM, Dorman BM (1986) Effect of
    mouse skin tumor promoters upon (3H)uridine exchange and focus
    formation in cultures of C3H10T1/2 mouse fibroblasts.
    Carcinogenesis, 7, 1095-1099

230

13. Chang CC, Trosko JE, Kung HJ, Bombick D, Matsumara F (1985) Potential role of the src gene product in inhibition of gap-junctional communication in NIH/3T3 cells. Proc. Natl. Acad. Sci. USA, 82, 5360-5364

14. Dotto GP, Weinberg RA, Ariza A (1988) Malignant transformation of mouse primary keratinocytes by Harvey sarcoma virus and its modulation by surrounding normal cells. Proc. Natl. Acad. Sci. USA, 85, 6389-6393

15. Elias PM, Grayson S, Caldwell TM, McNutt NS (1980) Gap junction proliferation in retinoic acid-treated human basal cell carcinoma. Lab. Invest., 42, 469-474

16. Enomoto K, Katoh F, Montesano R, Yamasaki H (1988) Oncogene alterations and intercellular communication in transformed and non-transformed rat liver epithelial cell lines. Proc. Amer. Assoc. Cancer Res., 29, 141

17. Enomoto T, Yamasaki H (1984) Lack of intercellular communication between chemically-transformed and surrounding non-transformed BALB/c 3T3 cells. Cancer Res., 44, 5200-5203

18. Enomoto T, Yamasaki H (1985) Phorbol ester-mediated inhibition of intercellular communication in BALB/c 3T3 cells: Relationship to enhancement of cell transformation. Cancer Res., 45, 2681-2688

19. Farr CJ, Saiki RK, Erlich HA, McCormick F, Marshall CJ (1988) Analysis of ras gene mutations in acute myeloid leukemia by polymerase chain reaction and oligonucleotide probes. Proc. Natl. Acad. Sci. USA, 85, 1629-1633

20. Hamel E, Katoh F, Mueller G, Birchmeier W, Yamasaki H (1988) Transforming growth factor β as a potent promoter in two-stage BALB/c 3T3 cell transformation. Cancer Res., 48, 2832-2836

21. Heimark RL, Schwartz SM (1985) The role of membrane-membrane interactions in the regulation of endothelial cell growth. J. Cell Biol., 100, 1934-1940

22. Hennings H, Jung R, Michael D, Yuspa SH (1986) An in vitro model for initiated mouse epidermis. Proc. Amer. Assoc. Cancer Res., 27, 1395

23. Herschman HR, Brankow DW (1986) Ultraviolet irradiation transforms C3H10T1/2 cells to a unique, suppressible phenotype. Science, 234, 1385-1388

24. Hirai H, Kobayashi Y, Mano H, Hagiwara K, Maru Y, Omine M, Mizoguchi H, Nishida J, Takaku F (1987) A point mutation at codon 13 of the N-ras oncogene in myelodysplastic syndrome. Nature, 327, 430-432

25. Kakunaga T, Hamada H, Learith J, Crow JD, Hirakawa T, Lo KY (1982) Evidence for both mutational and non-mutational process in chemically induced cell transformation. In: Harris CC, Cerutti PA, eds. Mechanisms of Chemical Carcinogenesis, Alan R. Liss, New York, pp. 517-529

26. Kalimi GH, Sirsat SM (1984) Phorbol ester tumor promoters affect the mouse epidermal gap junctions. Cancer Lett., 22, 343-350

27. Kam E, Pitts JD (1988) Effects of the tumor promoter 12-0-tetradecanoyl phorbol-13-acetate on junctional communication in intact mouse skin: Persistance of homologous communication and increase of epidermal-dermal coupling. Carcinogenesis, 9, 1389-1394

28. Klann RC, Fitzgerald DJ, Piccoli C, Slaga TJ, Yamasaki H (1989) Characterization of gap-junctional intercellular communication in SENCAR mouse epidermal cell lines. Cancer Res., 49, 699-705

29. Kraemer KH, Digiovanna JJ, Moshell AN, Tarone RE, Peck GL (1988) Prevention of skin cancer in xeroderma pigmentosum with the use of oral isotretinoin. New Engl. J. Med., 318, 1633-1637

30. Lo KY, Kakunaga T (1982) Similarities in the formation and removal of covalent DNA adduct in benzo(a)pyrene-treated BALB/c 3T3 variant cells with different induced transformation frequencies. Cancer Res., 42, 2644-2650

31. Matsukawa T, Bertram JS (1988) Augmentation of postconfluence growth arrest of 10T1/2 fibroblasts by endogenous cyclic adenosine 3':5'- monophosphate. Cancer Res., 48, 1874-1881

32. Mege RM, Matsuzaki F, Gallin WJ, Goldberg JI, Cunningham BA, Edelman GM (1988) Construction of epithelioid sheets by transfection of mouse sarcoma cells with cDNAs for chicken cell adhesion molecules. Proc. Natl. Acad. Sci. USA, 85, 7274-7278

33. Mehta PP, Bertram JS, Loewenstein WR (1986) Growth inhibition of transformed cells correlates with their junctional communication with normal cells. Cell, 44, 187-196

34. Mesnil M, Fitzgerald DJ, Yamasaki H (1988) Phenobarbital specifically reduces gap junction protein mRNA level in rat liver. Mol. Carcinogenesis, 1, 79-81

35. Mesnil M, Fraslin JM, Piccoli C, Yamasaki Y, Guguen-Guillouzo C (1987) Cell contact but not junctional communication (dye coupling) with biliary epithelial cells is required for hepatocytes to maintain differentiated functions. Exp. Cell Res., 173, 524-533

36. Mesnil M, Yamasaki H (1988) Selective gap junctional communication capacity of transformed and nontransformed rat-liver epithelial cell lines. Carcinogenesis, 9, 1499-1502

37. Milman HA, Elmore E, eds. (1987) Biochemical Mechanisms and Regulation of Intercellular Communication, Princeton Scientific Publishing Co., Princeton

38. Moolgavkar SH, Knudson AG Jr (1981) Mutation and cancer: a model for human carcinogenesis. J. Natl. Cancer Inst., 66, 1037-1052

39. Murray AW, Fitzgerald DJ (1979) Tumour promoters inhibit metabolic cooperation in cocultures of epidermal and 3T3 cells. Biochem. Biophys. Res. Commun., 91, 395-401

40. Pasti G, Rivedal E, Yuspa SH, Herald CL, Pettit GR, Blumberg PM (1988) Contrasting duration of cell-cell communication in primary mouse epidermal cells by phorbol 12,13-dibutyrate and by bryostatin 1. Cancer Res, 48, 447-451

41. Ponten J, MacIntyre EH (1968) Interaction between normal and transformed bovine fibroblasts in culture. II. Cells transformed by polyoma virus. J. Cell. Sci., 3, 603-613

42. Prutkin L (1975) Mucous metaplasia and gap junctions in the vitamin A acid-treated skin tumor, keratoacanthoma. Cancer Res., 35, 364-369

43. Quintanilla M, Brown K, Ramsden M, Balmain A (1986) Carcinogen-specific mutation and amplification of Ha-ras during mouse skin carcinogenesis. Nature, 322, 78-80

44. Rosen A, van der Merwe PA, Davidson JS (1988) Effects of SV40 transformation on intercellular gap junctional communication in human fibroblasts. Cancer Res., 48, 3485-3489

45. Sager R (1985) Genetic suppression of tumor formation. Adv. Cancer Res., 44, 43-68

46. Shay JW, Werbin H (1988) Cytoplasmic suppression of tumorigenicity in reconstructed mouse cells. Cancer Res., 48, 830-833

47. Sivak A, Van Duuren BL (1967) Phenotypic expression of transformation induction in cell culture by a phorbol ester. Science, 157, 1443-1444

48. Slaga TJ (1984) Can tumour promotion be effectively inhibited? In: Borzsonyi M, Lapis K., Day NE, Yamasaki H, eds. Models, Mechanisms and Etiology of Tumor Promotion, IARC Scientific Publications No. 56, IARC, Lyon, pp. 497-506

49. Stoker MGP (1967) Transfer of growth inhibition between normal and virus-transformed cells: Autoradiographic studies using marked cells. J. Cell. Sci., 2, 293-304

50. Stoker, MGP, Shearer M, O'Neill C (1966) Growth inhibition of polyoma- transformed cells by contact with static normal fibroblasts. J. Cell. Sci., 1, 297-310

51. Sugie S, Mori H, Takahashi M (1987) Effect of in-vivo exposure to the liver tumor promoters phenobarbital or DDT on the gap junctions of rat hepatocytes: a qualitative freeze-fracture analysis. Carcinogenesis, 8, 45-51

52. Takeda A, Hashimoto E, Yamamura H, Shimazu T (1987) Phosphorylation of liver gap junction protein by protein kinase C. FEBS Lett., 210, 169-172

53. Terzaghi-Howe M (1987) Inhibition of carcinogen-altered rat tracheal epithelial cell proliferation by normal epithelial cells in-vivo. Carcinogenesis, 8, 145-150

54. Vanhamme L, Rolin S, Szpirer C (1989) Inhibition of gap-junctional intercellular communication between epithelial cells transformed by the activated H-ras-1 oncogene. Exp. Cell Res., 180, 297-301

55. Volk T, Cohen O, Geiger B (1987) Formation of heterotypic adherens-type junctions between L-CAM-containing liver cells and A-CAM-containing lens cells. Cell, 50, 987-994

56. Wieser RJ, Oesch F (1986) Contact inhibition of growth of human diploid fibroblasts by immobilized plasma membrane glycoproteins. J. Cell Biol., 103, 361-367

57. Yamasaki H (1988) Role of gap-junctional intercellular communication in malignant cell transformation. In: Hertzberg EL, Johnson RG, eds. Gap Junctions (Modern Cell Biology, Vol. 7), Alan R. Liss Inc., New York, pp. 449-465

58. Yamasaki H (1989) Short-term assays to detect tumor-promoting activity of environmental chemicals. In: Slaga TJ, Klein-Szanto AJP, Boutwell RK, Stevenson DE, Spitzer HL, D'Motto B, eds., Skin Carcinogenesis: Mechanisms and Human Relevance, vol. 298, Alan R. Liss, Inc., New York, pp. 265-279

59. Yamasaki H, Enomoto T (1985) Role of intercellular communication in BALB/c 3T3 cell transformation. In: Barrett JC, Tennant RW, eds. Carcinogenesis, vol. 9, A Comprehensive Survey - Mammalian Cell Transformation: Mechanisms of Carcinogenesis and Assays for Carcinogens, Raven Press, New York, pp. 179-194

60. Yamasaki H, Enomoto K, Fitzgerald DJ, Mesnil M, Katoh F, Hollstein M (1988) Role of intercellular communication in the control of critical gene expression during multistage carcinogenesis. In: Kakunaga T, Sugimura T, Tomatis L, Yamasaki H, eds. Cell Differentiation, Genes and Cancer, IARC Scientific Publications No. 92, Lyon, pp. 57-75

61. Yamasaki H, Enomoto T, Martel N, Shiba Y, Kanno Y (1983) Tumor promoter-mediated reversible inhibition of cell-cell communication (electrical coupling): Relationship with phorbol ester binding and de novo macromolecule synthesis. Exp. Cell Res., 146, 297-308

62. Yamasaki H, Enomoto T, Shiba Y, Kanno Y, Kakunaga T (1985) Intercellular communication capacity as a possible determinant of transformation sensitivity of BALB/c 3T3 clonal cells. Cancer Res., 45, 637-641

63. Yamasaki H, Hollstein M, Martel N, Cabral JRP, Galendo D, Tomatis L (1987) Transplacental induction of a specific mutation in fetal Ha-ras and its critical role in post-natal carcinogenesis. Int. J. Cancer, 40, 818-822

64. Yamasaki H, Hollstein M, Mesnil M, Martel N, Aguelon AM (1987) Selective lack of intercellular communication between transformed and non-transformed cells. Cancer Res., 47, 5658-5664

65. Yamasaki H, Katoh F (1988) Further evidence for the involvement of gap junctional intercellular communication in induction and maintenance of transformed foci in BALB/c 3T3 cells. Cancer Res., 48, 3490-3495

66. Yotti LP, Chang CC, Trosko JE (1979) Elimination of metabolic cooperation in Chinese hamster cells by a tumor promoter. Science, 206, 1089-1091

INTERINDIVIDUAL VARIATION IN HUMAN CHEMICAL CARCINOGENESIS: IMPLICATIONS FOR RISK ASSESSMENT
Curtis C. Harris

Abstract

Chemical carcinogenesis is considered to be a multistage process involving genetic and epigenetic lesions that activate protooncogenes and inactivate tumor suppressor genes. Chemical carcinogens are generally activated enzymatically to electrophiles that form covalently bound carcinogen-DNA adducts. Detoxifying enzymes are competing with the activating enzymes for these procarcinogenic chemical substrates. Wide person to person variations in these two types of enzymatic activities are found. Repair rates of DNA damage caused by carcinogens also vary among individuals. These interindividual differences in the metabolism of chemical carcinogens and repair rates of carcinogen-induced DNA damage reflect acquired and inherited host factors that may influence an individual's risk for development of cancer. Interindividual and intertissue variation in response to tumor promoters is currently unknown in humans but studies in animal models would suggest substantial variability may occur. The wide interindividual variations in the determinants of genetic damage involved in tumor initiation and conversion and the assumed diversity in response to tumor promoters should be incorporated into efforts in quantitative assessment of cancer risk.

Introduction

"No one supposes that all the individuals of the same species are cast in the very same mould. These individual differences are highly important for us..." (Charles Darwin, The Origin of Species, 1859).

Scientists have repeatedly recognized person-to-person differences in behavior, morphology, and risk of disease. Such interindividual differences may be either acquired or inherited. For example, inherited differences in susceptibility to physical or chemical carcinogens have been observed among individuals, including an increased risk of sunlight-induced skin cancer in people with xeroderma pigmentosum [9], bladder cancer in dye stuff workers with a poor acetylator phenotype [7] and bronchogenic carcinoma in tobacco smokers who have an extensive debrisoquine hydroxylator phenotype [2].

236

Carcinogenesis is a multistage process involving both genetic
and epigenetic changes in the progenitor cells of cancer (Fig.1)
[3,4,26,64]. Because most chemical carcinogens require metabolic
activation to exert their oncogenic effects and the amount of
ultimate carcinogen produced results from the action of competing
activation and detoxication pathways, interindividual variation in
carcinogen metabolism is considered to be an important determinant of
cancer susceptibility. DNA adducts are one form of genetic damage
caused by chemical carcinogens and may lead to mutations that
activate proto-oncogenes and inactivate tumor suppressor genes in
replicating cells. The steady state levels of these adducts depend
on both the amount of ultimate carcinogens available and the rate of
removal from DNA by enzymatic repair processes. The genomic
distribution of adduct formation and repair is nonrandom and is
influenced by both DNA sequence and chromatin structure
[15,23,38,60].

This brief review will focus primarily on acquired and inherited
differences among individuals in both metabolism of chemical
carcinogens and DNA repair rates. The data have largely been
generated using *in vitro* models of human tissues, cells, or their
subcellular components [26]. Associations between interindividual
variations in these enzymatic activities and cancer susceptibility
will be emphasized. The likely variation to tumor promotion among
humans will also be discussed.

MULTISTAGE CARCINOGENESIS

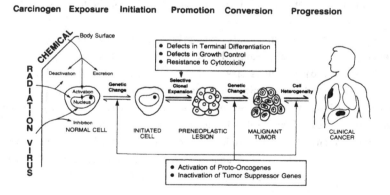

FIGURE 1   Simplistic model of multistage carcinogenesis

1. Carcinogen metabolism and DNA adduct formation.

Chemical carcinogens are metabolized by a wide variety of cytosolic and microsomal enzymes. Multiple forms of human cytochrome P_{450} are involved in the oxidative metabolism of chemical carcinogens, e.g., polycyclic aromatic hydrocarbons [22]. Several thousand-fold interindividual variation has been observed in placental aryl hydrocarbon hydroxylase (AHH) activity; some of this variability is under direct genetic control, but variations are also the result of an enzyme induction process due to maternal exposure to environmental carcinogens, e.g., tobacco smoke or dietary factors [10,18,37]. However, the induction process itself may have a genetic component [41] and inducible AHH activity is higher in cultured lymphocytes from lung cancer cases when compared to controls [32]. Trell and coworkers [61] are currently conducting a prospective study to determine if high inducibility of AHH is a risk factor for the development of cancer. Increased binding of benzo[a]pyrene metabolites to DNA in lymphocytes of patients with lung cancer versus non-cancer controls has also been observed [28].

Cytochrome $P_{450IID1}$ activity is polymorphic and has also been linked to lung cancer risk [2]. This P_{450} hydroxylates xenobiotics such as debrisoquine and an individual's polymorphic phenotype is inherited in an autosomal recessive manner. The 4-hydroxylation of debrisoquine varies several thousand-fold among people, and lung, liver, or advanced bladder cancer patients are more likely to have the extensive hydroxylator phenotype when compared to noncancer controls [2,29,31]. In a case-control study of lung cancer in the U.S. the extensive hydroxylator phenotype had a 14-fold increased cancer risk when compared to the poor hydroxylator phenotype and the increased risk was primarily for histological types other than adenocarcinoma of the lung [6]. In addition, further analysis of the British data [2] has revealed that individuals with the extensive hydroxylator phenotype who are occupationally exposed to high amounts of either asbestos or polycyclic aromatic hydrocarbons have relative lung cancer risks of 18- and 35-fold, respectively [5]. Although it is attractive to speculate that $P_{450IID1}$ may activate a chemical carcinogen(s) found in tobacco smoke, thereby rendering an individual with the extensive metabolizor phenotype at greater risk, carcinogenic substrates for this cytochrome $P_{450IID1}$ have not as yet been identified. An alternative but less attractive hypothesis is that the $P_{450IID1}$ gene is in linkage disequilibrium with a different gene that does influence cancer susceptibility.

The N-acetylation polymorphism is controlled by two autosomal alleles at a single locus in which rapid acetylation is the dominant trait and slow acetylation is recessive. Acetylation of carcinogenic aromatic amines has been proposed as a cancer risk factor. The slow acetylator phenotype has been linked to occupationally-induced bladder cancer in dye workers exposed to large amounts of N-substituted aryl compounds [7]. Interestingly, the rapid acetylator phenotype is commonly found in colon cancer cases [34]. Whether or not this association is due to metabolism of a carcinogenic aromatic amine in the colonic epithelium is not known.

Wide interindividual differences in detoxifying enzymes of carcinogens are also found. For example, at each step in the metabolic pathway of benzo[a]pyrene activation to electrophilic diol epoxides competing detoxifying enzymes are found. A recent study of several of the enzymes involved in benzo[a]pyrene metabolism showed a more than 10-fold person-to-person variation in their activities [48] and presented indirect evidence that tobacco smoke induced many of these enzymes. Genetic control of the presumed detoxication of benzo[a]pyrene by conversion to water-soluble metabolites has been reported [41].

Metabolism of carcinogens from several chemical classes, including N-nitrosamines, polycyclic aromatic hydrocarbons, hydrazines, mycotoxins, and aromatic amines, has been studied in cultured human tissues and cells [1,13,14,24,39,43,55,66]. The enzymes responsible for the activation and deactivation of procarcinogens, the metabolites produced, and the carcinogen-DNA adducts formed by cultured human tissues and cells, in general, are qualitatively similar among donors and tissue types. The DNA adducts and carcinogen metabolites are also similar to those found in most laboratory animals, an observation that supports the qualitative extrapolation of carcinogenesis data from the laboratory animal to the human. Some notable differences among animal species have been reported, including metabolism of aromatic amines in the guinea pig [19], benzo[a]pyrene in the rat [59], and aflatoxin B_1 in the Syrian golden hamster [58].

Although the major DNA adducts are qualitatively similar for the chemical carcinogens so far studied in these *in vitro* models, quantitative differences have been found among individuals and their various tissue types [1,13,33,44,62]. The differences in formation of DNA adducts range from approximately 10- to 150-fold among humans, the interindividual distribution is generally unimodal, and the variation is similar in magnitude to that found in pharmacogenetic studies of drug metabolism.

2. DNA repair rates.

DNA repair enzymes modify DNA damage caused by carcinogens, including removal of DNA adducts. Studies of cells from donors with xeroderma pigmentosum have been particularly important in expanding our understanding of DNA excision repair and its possible relationship to risk of cancer [9]. The rate but not the fidelity of DNA repair can be determined by measuring unscheduled DNA synthesis and removal of DNA adducts, and substantial interindividual variations in DNA repair rates have been observed [54]. In addition to finding excision repair rates severely depressed in xeroderma pigmentosum cells (*e.g.*, complementation group A), an approximately 5-fold variation among individuals in unscheduled DNA synthesis induced by UV exposure of lymphocytes *in vitro* has been found in the general population [54]. A significant reduction in unscheduled DNA synthesis induced *in vitro* by N-acetoxy-2-acetylaminofluorene has been observed in mononuclear leukocytes from individuals with a history of cancer in first-degree relatives when compared to those without a family history [46,47].

Interindividual variation has been noted in the activity of O^6-alkylguanine-DNA alkyltransferase; this enzyme repairs alkylation damage to O^6-deoxyguanine [11,12,20,35]. In addition to these person-to-person differences, wide variations in this DNA repair activity have been observed in different types of tissues [11,12,21,65], and fetal tissues exhibit 2- to 5-fold less activity than the corresponding adult tissues [40], and cells may have lower repair rates after terminal differentiation [30]. The activity of this DNA repair enzyme is inhibited by certain aldehydes [21] and alkylating cancer chemotherapeutic agents [53]. A decrease in this DNA repair activity has been observed in fibroblasts from patients with lung cancer when compared to donors with either melanoma or noncancer controls [52]. Therefore, acquired and/or inherited deficiency in 0^6-alkylguanine-DNA-alkyltransferase may be a cancer risk factor in tobacco smokers.

An unimodal distribution of repair rates of benzo[a]pyrene diol epoxide-DNA adducts has been observed using human lymphocytes *in vitro* [42]. The interindividual variation was substantially greater than the intraindividual variation which suggests the influence of inherited factors. The influence of these variations in DNA repair rates in determining tissue site and risk of cancer in the general population remains to be determined.

3. Prospective and future studies.

The interindividual variations found in carcinogen metabolism, DNA adduct formation and DNA repair rates are influenced by a complex interaction between acquired and inherited host factors. Therefore, the contribution of inheritance to these activities and their importance in cancer susceptibility is difficult to determine. However, rapid advances are being made in: (a) cloning the cytochrome P450 genes involved in carcinogen metabolism; (b) determining the molecular control of their expression by positive and negative regulatory elements, that may be constitutive as well as inducible; (c) defining mutations such as deletions involved in genetic polymorphism; (d) isolation of their gene products; and (e) the development of immunoassays to measure them [22]. More than 15 types of human P_{450} have been genetically cloned [22]. In one case, i.e., $P_{450IID1}$, DNA restriction fragment length polymorphism (RFLP) analysis can be used to identify about 75% of the individuals with the poor hydroxylator phenotype [56]. Another promising approach is to isolate mRNA from normal lymphocytes, reverse transcribe the mRNA into cDNA derivatives, amplify these derivatives by the polymerase chain reaction (PCR), and determine their size by electrophoretic analysis. A preliminary study [51] has shown that the sizes of the PCR products are indicative of an individual's phenotype. Investigators are searching for RFLPs that are predictive of oxidative propensity of the various forms of the P_{450} in the multigene family so that an individual's phenotype can be tested at the DNA level without the current necessity to administer xenobiotics to people. If this approach proves to be feasible, an individual could be genotyped by RFLP analysis *in utero* or at birth and, if necessary, steps could be taken to prevent exposure to certain chemicals and to utilize other types of intervention.

Efforts to clone DNA repair genes have proven to be difficult in part because of the complex nature of these genes. For example, the human DNA repair gene, ERCC-1, is an evolutionary composite of the *uvr* A and C genes found in bacteria and rad 10 gene found in yeast [63]. As human DNA repair genes are isolated, the fidelity of these enzymes in the repair process can be determined and more specific and rapid assays can be developed so that the role of interindividual variation in DNA repair during chemical carcinogenesis can be better understood.

Because cancer is the result of complex interaction between multiple environmental factors and both acquired and inherited host factors [25], one should consider carcinogen-DNA adducts as only one piece in the puzzle. Examples of other portions of the puzzle include determinants of tumor promotion. In the skin carcinogenesis studies, wide variations in susceptibility to tumor-promoting agents have been observed among animal species and among different inbred strains of a single species (Table 1) [57]. Epidemiological studies suggest that tumor promotion may influence both tumor probability and latency period in humans [16]. Methods to predict responses to tumor promoters among humans need to be developed.

TABLE 1: Sensitivity to Skin Carcinogenesis in Different Stocks and Strains of Mice

	Sensitivity[a]
Complete carcinogenesis	Sencar>CD-1>C57BL/6≥ICR/Ha Swiss>C3H
Two-stage carcinogenesis (initiation-promotion)	Sencar>>CD-1>ICR/Ha Swiss≥BALB/c>C57BL/6≥C3H≥DBA/2

[a] Data represent sensitivities of various mouse strains to benzo(a)pyrene and 7,12-dimethylbenz(a)anthracene. Ranking represents a subjective view because dose-response data were not available for all strains [25].

There is also an increasing amount of data which suggests that chemical carcinogens may cause both direct DNA damage, i.e., carcinogen-DNA adducts, and indirect DNA damage by causing formation of free radicals and superoxides that react with DNA and cause molecular lesions, e.g., thymine glycol [8]. Carcinogens can damage membranes and initiate the arachidonic acid cascade and the release of lipid peroxidation aldehydes, such as 4-hydroxyalkenals [17] that bind to DNA. Phthalates and hypolipidemic drugs, including clofibrate, apparently act through an indirect mechanism by causing proliferation of peroxisomes and a subsequent increase in superoxides [50]. Measures of indirect DNA damage are needed, e.g., the development of monoclonal antibodies to thymine glycol in DNA [36] and ^{32}P-postlabeling assay of 8-OH-deoxyguanine [49].

The primary goal of molecular epidemiology is to identify individuals at increased cancer risk by obtaining evidence of high exposure to carcinogens, leading to pathobiological lesions in target cells, and/or increased oncogenic susceptibility due to either inherited or acquired host factors. This multidisciplinary area of cancer research combines epidemiological and laboratory approaches [27,45]. Interindividual variations among humans in carcinogen biodistribution and metabolism, DNA adduct formation and DNA repair have important implications in molecular epidemiology of cancer risk. An increased understanding of the molecular basis of these differences among humans and their linkage with critical steps in carcinogenesis may afford in the future the potential to predict disease risk for individual persons, instead of populations, and before the onset of clinically evident disease.

4. Implications of interindividual variations in cancer risk assessment.

An individual's cancer risk is dependent on interactive effects of exposure to carcinogens and host factors that influence susceptibility. This brief review has focused primarily on acquired and inherited host factors and their individual-to-individual variability. Individual differences in exposure to carcinogens also influence one's cancer risk.

Examples of generic uncertainties in the full utilization of this information on host factors in mechanistic-physiological modeling for the purpose of quantitative cancer risk assessment are listed in Table 2. Since the contribution of each type of host risk factor in determining an individual's risk will vary, a potency ranking or index is needed. For example, the activites of rate-limiting enzymes in either the activation or detoxification of chemical carcinogens should carry more statistical weight than non-rate limiting enzymes involved in carcinogen metabolism. The efficiency of the mucociliary transport system in the clearance of inhaled particulates is more likely to be a risk factor for asbestos and carcinogens bound to particulates than for gaseous carcinogens such as formaldehyde.

It follows from the above discussion that interindividual variation in host factors with a higher impact on risk will generally be more important than the variation in host factors of lesser importance. A simple illustration of the importance of interindividual variation can be drawn from two host factors that influence an individual's risk of skin cancer. The lighter the color of one's skin, the greater the risk of certain types of skin cancer in individuals exposed to equivalent amounts of sunlight. However, this inherited host factor is much less important in an individual with xeroderma pigmentosum in which a severe deficiency in repair of DNA damage caused by ultraviolet light leads to increased sensitivity to the carcinogenicity of sunlight. Xeroderma pigmentosum is also an obvious example of an extreme case of genetic disposition.

TABLE 2: Interindividual Variation and Risk Assessment: Examples of Generic Uncertainties

---

1. Rank order of "potency or weight" of various host susceptibility factors influencing the cancer risk in individuals exposed to carcinogens
2. "Potency" impact of the degree of the interindividual variation in a specific host susceptibility factor
3. Interactive effects among various host susceptibility factors

---

Xeroderma pigmentosum is the classic example of an inherited host susceptibility factor at one extreme of the spectrum. However, the distribution of individuals with varying deficiencies in DNA repair of the incision type as well as other host susceptibility factors needs to be determined and considered in the formulation of mathematical models of risk assessment (see paper by Portier *et al.*, this volume).

Another example of the interactive effects of exposure to increasing amounts of environmental carcinogens and an inherited host factor is shown in Table 3. Increasing estimated exposure levels of asbestos or polycyclic aromatic hydrocarbons may have more than additive effects in enhancing the relative risk of bronchogenic carcinoma in people with the extensive debrisoquine phenotype.

Biomedical scientists will continue to identify host susceptibility factors, determine their interindividual variation and understand their relative contribution to human carcinogenesis. The incorporation of current and future information concerning interindividual variations in acquired and inherited host susceptibility factors in mathematical models generating quantitative risk assessments will improve the quality and scientific credibility of the estimates.

TABLE 3: Increased Relative Risk of Lung Cancer due to Integration of Occupational Carcinogen Exposure to Asbestos or Polycyclic Aromatic Hydrocarbon (PAH) and the Extensive Metabolizer Debrisoquine Metabolic Phenotype [5]

	Asbestos	PAH
• Relative risk due to occupational exposure in low risk debrisoquine metabolic phenotype	1.8	0.7
• Relative risk in occupationally unexposed and high risk debrisoquine metabolic phenotype	6.0	3.9
• Relative risk in occupationally exposed and high risk debrisoquine metabolic group	18.4	35.3

"In biology, one rarely deals with classes of identical
entities, but nearly always studies populations consisting of unique
individuals. This is true for every level of hierarchy, from cells
to ecosystems" (Ernst Mayr, The Growth of Biological Thought, 1982).

Acknowledgment

The editorial aid of Robert Julia is appreciated.

References

1. Autrup, H., and Harris, C.C. Metabolism of chemical carcinogens
   by cultured human tissues. In: C.C. Harris and H. Autrup (eds.),
   Human Carcinogenesis, pp. 169-194. New York: Academic Press,
   1983.

2. Ayesh, R., Idle, J.R., Ritchie, J.C., Crothers, M.J., and Hetzel,
   M.R. Metabolic oxidation phenotypes as markers for
   susceptibility to lung cancer. Nature, 312: 169-170, 1984.

3. Barbacid, M. Oncogenes and human cancer: cause or consequence.
   Carcinogenesis, 7: 1037-1042, 1986.

4. Bishop, J.M. The molecular genetics of cancer: 1988. Leukemia, 2:
   199-208, 1988.

5. Caporaso, N., Hayes, R.B., Dosemeci, M., Hoover, R., Ayesh, R.,
   Hetzel, M., and Idle, J.R. Lung cancer risk, occupational
   exposure, and the debrisoquine metabolic phenotype. Cancer Res.,
   49: 3675-3679, 1989.

6. Caporaso, N., Hoover, R., Eisner, S., Resau, J., Trump, B.F.,
   Issaq, H., Morshik, G., and Harris, C.C. Debrisoquine (D)
   metabolic phenotype (MP) and the risk of lung cancer. ASCO
   Proceedings, 7: 336-336, 1988.(Abstract)

7. Cartwright, R.A., Glashan, R.W., Rogers, H.J., Ahmad, R.A.,
   Barham-Hall, D., Higgins, E., and Kahn, M.A. Role of
   N-acetyltransferase phenotypes in bladder carcinogenesis: a
   pharmacogenetic epidemiological approach to bladder cancer.
   Lancet, 2: 842-845, 1982.

8. Cerutti, P.A. Prooxidant states and tumor promotion. Science,
   227: 375-381, 1985.

9. Cleaver, J.D., Bodell, W.J., Gruenert, D.C., Kapp, L.N., Kaufman,
   W.K., Park, S.D., and Zelle, B. Repair and replication
   abnormalities in various human hypersensitive diseases. In: C.C.
   Harris and P.A. Cerutti (eds.), Mechanisms of Chemical
   Carcinogenesis, pp. 409-416. New York: Alan R.Liss, 1982.

10. Conney, A.H. Induction of microsomal enzymes by foreign chemicals and carcinogenesis by polycyclic aromatic hydrocarbons: G. H. A Clowes Memorial Lecture. Cancer Res., $\underline{42}$: 4875-4917, 1982.

11. D'Ambrosio, S.M., Samuel, M.J., Dutta-Choudhury, T.A., and Wani, A.A. O6-methylguanine-DNA methyltransferase in human fetal tissues: fetal and maternal factors. Cancer Res., $\underline{47}$: 51-55, 1987.

12. D'Ambrosio, S.M., Wani, G., Samuel, M., and Gibson-D'Ambrosio, R.E. Repair of O6-methylguanine in human fetal brain and skin cells in culture. Carcinogenesis, $\underline{5}$: 1657-1661, 1984.

13. Daniel, F.B., Stoner, G.D., and Schut, H.A.J. Interindividual variation in the DNA binding of chemical genotoxins following metabolism by human bladder and bronchus explants. In: F.J. DeSerres and R. Pero (eds.), Individual Susceptibility to Genotoxic Agents in Human Population, pp. 177-199. New York: Plenum Press, 1983.

14. De Flora, S., Petruzzelli, S., Camoirano, A., Bennicelli, C., Romano, M., Rindi, M., Ghelarducci, L., and Giuntini, C. Pulmonary metabolism of mutagens and its relationship with lung cancer and smoking habits. Cancer Res., $\underline{47}$: 4740-4745, 1987.

15. Dolan, M.E., Oplinger, M., and Pegg, A.E. Sequence specificity of guanine alkylation and repair. Carcinogenesis, $\underline{9}$: 2139-2143, 1988.

16. Doll, R., and Peto, R. The Causes of Cancer. New York: Oxford Press, 1981.

17. Esterbauer, H. Aldehydic products of lipid peroxidation. In: D.C.H. McBrien and T.F. Slater (eds.), Free Radicals, Lipid Peroxidation and Cancer, pp. 101-158. New York: Academic Press, 1982.

18. Fujino, T., Gottlieb, K., Manchester, D.K., Park, S.S., West, D., Gurtoo, H.L., Tarone, R.E., and Gelboin, H.V. Monoclonal antibody phenotyping of interindividual differences in cytochrome P-450-dependent reactions of single and twin human placenta. Cancer Res., $\underline{44}$: 3916-3923, 1984.

19. Garner, R.C., Martin, C.N., and Clayson, D.B. Carcinogenic aromatic amines and related compounds in chemical carcinogens. In: A. Searle (ed.), Chemical Carcinogens, pp. 175-276. Washington,DC: American Chemical Society, 1984.

20. Gerson, S.L., Miller, K., and Berger, N.A. O6 alkylguanine-DNA alkyltransferase activity in human myeloid cells. J.Clin.Invest., $\underline{76}$: 2106-2114, 1985.

21. Grafstrom, R.C., Pegg, A.E., Trump, B.F., and Harris, C.C. O6-alkylguanine-DNA alkyltransferase activity in normal human tissues and cells. Cancer Res., 44: 2855-2857, 1984.

22. Guengerich, F.P. Characterization of human microsomal cytochrome P-450 enzymes. Annu.Rev.Pharmacol.Toxicol., 29: 241-264, 1989.

23. Hanawalt, P.C. Preferential DNA repair in expressed genes. Environ.Health Perspect., 76: 9-14, 1987.

24. Harries, G.C., Boobis, A.R., Collier, N., and Davies, D.S. Interindividual differences in the activation of two hepatic carcinogens to mutagens by human liver. Hum.Toxicol., 5: 21-26, 1986.

25. Harris, C.C. Concluding remarks: Role of carcinogens, cocarcinogens and host factors in cancer risk. In: C.C. Harris and H. Autrup (eds.), Human Carcinogenesis, pp. 941-970. New York: Academic Press, 1983.

26. Harris, C.C. Human tissues and cells in carcinogenesis research. Cancer Res., 47: 1-10, 1987.

27. Harris, C.C., Weston, A., Willey, J.C., Trivers, G.E., and Mann, D.L. Biochemical and molecular epidemiology of human cancer: indicators of carcinogen exposure, DNA damage, and genetic predisposition. Environ.Health Perspect., 75: 109-119, 1987.

28. Hawke, L.J., and Farrell, G.C. Increased binding of benzo[a]pyrene metabolites to lymphocytes from patients with lung cancer. Cancer Lett., 30: 289-297, 1986.

29. Idle, J.R., Mahgoub, A., Sloan, T.P., Smith, R.L., Mbanefo, C.O., and Bababunmi, E.A. Some observations on the oxidation phenotype status of Nigerian patients presenting with cancer. Cancer Lett., 11: 331-338, 1981.

30. Jensen, L., and Linn, S. A reduced rate of bulky DNA adduct removal is coincident with differentiation of human neuroblastoma cells induced by nerve growth factor. Mol.Cell Biol., 8: 3964-3968, 1988.

31. Kaisary, A., Smith, P., Jaczq, E., McAllister, C.B., Wilkinson, G.R., Ray, W.A., and Branch, R.A. Genetic predisposition to bladder cancer: ability to hydroxylate debrisoquine and mephenytoin as risk factors. Cancer Res., 47: 5488-5493, 1987.

32. Kouri, R.E., McKinney, C.E., Slomiany, D.J., Snodgrass, D.R., Wray, N.P., and McLemore, T.L. Positive correlation between high aryl hydrocarbon hydroxylase activity and primary lung cancer as analyzed in cryopreserved lymphocytes. Cancer Res., 42: 5030-5037, 1982.

33. Kuroki, T., Hosomi, J., Chida, K., Hosoi, J., and Nemoto, N. Inter-individual variations of arylhydrocarbon-hydroxylase activity in cultured human epidermal and dermal cells. Jpn.J.Cancer Res., 78: 45-53, 1987.

34. Lang, N.P., Chu, D.Z., Hunter, C.F., Kendall, D.C., Flammang, T.J., and Kadlubar, F.F. Role of aromatic amine acetyltransferase in human colorectal cancer. Arch.Surg., 121: 1259-1261, 1986.

35. Lawley, P.D., Harris, G., Phillips, E., Irving, W., Colaco, C.B., Lydyard, P.M., and Roitt, I.M. Repair of chemical carcinogen-induced damage in DNA of human lymphocytes and lymphoid cell lines--studies of the kinetics of removal of O6-methylguanine and 3-methyladenine. Chem.Biol.Interact., 57: 107-121, 1986.

36. Leadon, S.A., and Hanawalt, P.C. Monoclonal antibody to DNA containing thymine glycol. Mutat.Res., 112: 191-200, 1983.

37. Manchester, D.K., and Jacoby, E.H. Sensitivity of human placental monooxygenase activity to maternal smoking. Clin.Pharmacol.Ther., 30: 687-692, 1981.

38. Mattes, W.B., Hartley, J.A., Kohn, K.W., and Matheson, D.W. GC-rich regions in genomes as targets for DNA alkylation. Carcinogenesis, 9: 2065-2072, 1988.

39. Minchin, R.F., McManus, M.E., Boobis, A.R., Davies, D.S., and Thorgeirsson, S.S. Polymorphic metabolism of the carcinogen 2-acetylaminofluorene in human liver microsomes. Carcinogenesis, 6: 1721-1724, 1985.

40. Myrnes, B., Giercksky, K.E., and Krokan, H. Interindividual variation in the activity of O6-methyl guanine-DNA methyltransferase and uracil-DNA glycosylase in human organs. Carcinogenesis, 4: 1565-1568, 1983.

41. Nowak, D., Schmidt-Preuss, U., Jorres, R., Liebke, F., and Rudiger, H.W. Formation of DNA adducts and water-soluble metabolites of benzo[a]pyrene in human monocytes is genetically controlled. Int.J.Cancer, 41: 169-173, 1988.

42. Oesch, F., Aulmann, W., Platt, K.L., and Doerjer, G. Individual differences in DNA repair capacities in man. Arch.Toxicol.[Suppl], 10: 172-179, 1987.

43. Pelkonen, O., and Nebert, D.W. Metabolism of polycyclic aromatic hydrocarbons: etiologic role in carcinogenesis. Pharmacol.Rev., 34: 189-222, 1982.

44. Perera, F.P., Santella, R.M., Brenner, D., Poirier, M.C., Munshi, A.A., Fischman, H.K., and Van Ryzin, J. DNA adducts, protein adducts, and sister chromatid exchange in cigarette smokers and nonsmokers. JNCI, 79: 449-456, 1987.

45. Perera, F.P., and Weinstein, I.B. Molecular epidemiology and carcinogen-DNA adduct detection: new approaches to studies of human cancer causation. J.Chronic.Dis., 35: 581-600, 1982.

46. Pero, R.W., Bryngelsson, C., Bryngelsson, T., and Norden, A. A genetic component of the variance of N-acetoxy-2-acetylaminofluorene-induced DNA damage in mononuclear leukocytes determined by a twin study. Hum.Genet., 65: 181-184, 1983.

47. Pero, R.W., Johnson, D.B., Markowitz, M., Doyle, G., Lund-Pero, M., Seidegard, J., Halper, M., and Miller, D.G. DNA repair synthesis in individuals with and without a familial history of cancer. Carcinogenesis, 1988.in press

48. Petruzzelli, S., Camus, A.M., Carrozzi, L., Ghelarducci, L., Rindi, M., Menconi, G., Angeletti, C.A., Ahotupa, M., Hietanen, E., Aitio, A., Saracci, R., Bartsch, H., and Giuntini, C. Long-lasting effects of tobacco smoking on pulmonary drug-metabolizing enzymes: a case-control study on lung cancer patients. Cancer Res., 48: 4695-4700, 1988.

49. Povey, A.C., Wilson, V.L., Taffe, B.G., Wood, M.L., Essigmann, J.M., and Harris, C.C. Detection and quantitation of 8-hydroxydeoxyguanosine residues by 32P-postlabelling. Proc.Am.Assoc.Cancer Res., 30: 201-201, 1989.(Abstract)

50. Reddy, J.K., and Lalwani, N.D. Carcinogenesis by hepatic peroxisome proliferators: evaluation of the risk of hypolipodemic drugs and industrial plasticizers to humans. CRC Crit.Rev.Toxicol., 12: 1-58, 1984.

51. Redlich, C.A., Juilfs, D.M., and Omiecinski, C.J. Genetic expression of the debrisoquine 4-hydroxylase polymorphism in human lymphocytes. Clin.Res., 37: 341A-341A, 1989.(Abstract)

52. Rudiger, H.W., Schwartz, U., Serrand, E., Stief, M., Krause, T., Nowak, D., Doerjer, G., and Lehnert, G. Reduced O6-methylguanine repair in fibroblast cultures from patients with lung cancer. Carcinogenesis, 1988.in press

53. Sagher, D., Karrison, T., Schwartz, J.L., Larson, R., Meier, P., and Strauss, B. Low O6-alkylguanine DNA alkyltransferase activity in the peripheral blood lymphocytes of patients with therapy-related acute nonlymphocytic leukemia. Cancer Res., 48: 3084-3089, 1988.

54. Setlow, R.B. Variations in DNA repair among humans. In: C.C. Harris and H. Autrup (eds.), Human Carcinogenesis, pp. 231-254. New York: Academic Press, 1983.

55. Siegfried, J.M., Rudo, K., Bryant, B.J., Ellis, S., Mass, M.J., and Nesnow, S. Metabolism of benzo(a)pyrene in monolayer cultures of human bronchial epithelial cells from a series of donors. Cancer Res., 46: 4368-4371, 1986.

56. Skoda, R.C., Gonzalez, F.J., Demierre, A., and Meyer, U.A. Two mutant alleles of the human cytochrome P-450db1 gene (P450C2D1) associated with genetically deficient metabolism of debrisoquine and other drugs. Proc.Natl.Acad.Sci.USA., 85: 5240-5243, 1988.

57. Slaga, T.J., Fischer, S.M., Weeks, C.E., Klein-Szanto, A.J., and Reiners, J.J. Studies of mechanisms involved in multistage carcinogenesis in mouse skin. In: C.C. Harris and P.A. Cerutti (eds.), Mechanisms of Chemical Carcinogenesis, pp. 207-227. New York: Alan R. Liss, Inc., 1982.

58. Stoner, G.D., Daniel, F.B., Schenck, K.M., Schut, H.A., Sandwisch, D.W., and Gohara, A.F. DNA binding and adduct formation of aflatoxin B1 in cultured human and animal tracheobronchial and bladder tissues. Carcinogenesis, 3: 1345-1348, 1982.

59. Stowers, S.J., and Anderson, M.W. Formation and persistence of benzo(a)pyrene metabolite- DNA adducts. Environ.Health Perspect., 62: 31-39, 1985.

60. Topal, M.D. DNA repair, oncogenes and carcinogenesis. Carcinogenesis, 9: 691-696, 1988.

61. Trell, L., Korsgaard, R., Janzon, L., and Trell, E. Distribution and reproducibility of aryl hydrocarbon hydroxylase inducibility in a prospective population study of middle- aged male smokers and nonsmokers. Cancer, 56: 1988-1994, 1985.

62. Vahakangas, K., Autrup, H., and Harris, C.C. Interindividual variation in carcinogen metabolism, DNA damage and DNA repair. In: A. Berlin, M. Draper, K. Hemminki and H. Vainio (eds.), Monitoring Human Exposure to Carcinogenic and Mutagenic Agents, pp. 85-98. Lyon: IARC Sci.Publ., 1984.

63. Van Duin, M., de Wit, J., Odijk, H., Westerveld, A., Yasui, A., Koken, H.M., Hoeijmakers, J.H., and Bootsma, D. Molecular characterization of the human excision repair gene ERCC-1: cDNA cloning and amino acid homology with the yeast DNA repair gene RAD10. Cell, 44: 913-923, 1986.

64. Weinstein, I.B. The origins of human cancer: molecular mechanisms of carcinogenesis and their implications for cancer prevention and treatment--twenty-seventh G.H.A. Clowes memorial award lecture. Cancer Res., 48: 4135-4143, 1988.

65. Yarosh, D.B. The role of O6-methylguanine-DNA methyltransferase in cell survival, mutagenesis and carcinogenesis. Mutat.Res., <u>145</u>: 1-16, 1985.

66. Yoo, J.S., Guengerich, F.P., and Yang, C.S. Metabolism of N-nitrosodialkylamines by human liver microsomes. Cancer Res., <u>48</u>: 1499-1504, 1988.

UTILIZING BIOLOGICALLY BASED MODELS TO ESTIMATE CARCINOGENIC RISK

Christopher J. Portier

ABSTRACT: The limitations associated with using biologically-based mathematical models for the estimation of carcinogenic risks from long-term chemical exposures at low dose levels represents a statistical and mathematical challenge with special relevance to environmental research. Determining an adequate model for estimating the relationship between dose and response is critical to reducing potential bias in the risk estimation process. This talk discusses the various assumptions and models used in carcinogenic risk assessment. The emphasis will be on the accuracy with which the magnitude of the carcinogenic risk, the shape of the dose-response relationship and the overall variability of the risk estimates can be determined from the available data.

1. Estimation of Carcinogenic Risk

The use of animal models in estimating the carcinogenic potential of a chemical in humans assumes that there exists some measure of dose, termed by the EPA (1986) as an equivalent dose (ED), for which the rate of tumor onset in the test species is equivalent to the rate of tumor onset in humans. This assumption implies there are two distinct classes of models which are used to relate carcinogenicity in animals to carcinogenicity in humans. The first class of models are referred to as "species extrapolation models" and relate the external exposure level of the compound in animals or in humans to the assumed ED. These models account for differences in the production of the ED across species. The ED may be the parent compound, some metabolite of the parent compound, some other derivative of the compound or a combination of these. The second class of models are referred to as "tumor incidence models". These relate the ED of the chemical to the probability of cancer. This class of models includes statistical models, empirical mathematical models and "mechanistic" models that characterize the hypothesized mechanism through which normal cells become malignant cells.

The tumor incidence model is used to estimate a safe ED. The safe ED is defined as the equivalent dose that yields a pre-specified negligible increase in carcinogenic risk over the spontaneous risk. With the initial assumption concerning an equivalent dose across species, this safe ED in animals represents a safe ED for humans. The safe ED for humans is then converted into a safe human exposure level via a model which relates ED to D for humans (usually the same conceptual model that related D to ED for the animal species but with parameters relevant to humans). The basic steps just described for estimating a safe exposure level are presented as an outline in Table 1. This description pertains to the actual numerical determination of a risk estimate. The setting of exposure standards involves a variety of disciplines besides the biological and mathematical sciences. Many aspects of the process of determining exposure standards are subjective and, as such, are not quantifiable. In what follows, some of the critical issues of the procedure outlined in Table 1 for carcinogenic risk estimation will be discussed in greater detail.

TABLE 1: Basic Steps In Estimating A Safe Exposure Level

1. Consider Probable Mechanisms
    A. Utilize available knowledge to determine an equivalent dose (ED)
    B. Determine an appropriate tumor incidence model
2. Estimate the relationship between the administered dose and equivalent dose in the experimental animal
3. Express the administered doses in the carcinogenicity experiment in units of the equivalent dose
4. Estimate parameters of the tumor incidence model as a function of the equivalent dose
5. Determine a safe equivalent dose level
6. Estimate the relationship between the equivalent dose and the exposure dose in humans.
7. Express the safe equivalent dose in units of exposure dose in humans.

---

2. Biologically-Based Tumor Incidence Models

Many tumor incidence models have been proposed for estimating carcinogenic risk and most do an adequate job of fitting the animal tumor incidence data in the experimental dose range. These models can differ considerably in the low-dose region, which is usually the critical region for estimating a safe ED. This is illustrated by Figure 1. In Part A of Figure 1, two different tumor incidence models are plotted along with a hypothetical set of animal tumor incidence data. It is clear that both of these models adequately describe this data. If a model is chosen for which the slope of the dose-response curve is positive at dose zero, then small increases in dose will result in proportionate increases in the risk of getting a tumor. Models of this type are referred to as "low-dose linear" models and are illustrated by the "Linear Model" in Figure 1. Models for which the slope of the dose-

254

response curve is zero at dose zero would result in virtually no change in carcinogenic risk for small dosage levels. These models are typically referred to as "non-linear" models and the "Non-Linear Model" shown in Figure 1 is of this type.

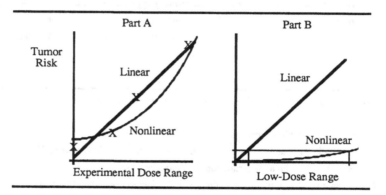

FIGURE 1: Linear and Non-Linear Dose-Response Curves in the
Observable Response Range

It is often difficult to determine which model is appropriate for a given set of data in the experimental dose range. Yet, if we examine added-risk over background and focus on the low-dose region of these dose-response curves (detailed in Part B of Figure 1), there is a dramatic difference in the two curves. The "Linear Model" predicts a substantially larger risk for a given small dose than does the "Non-Linear Model".

The implications of an incorrect model choice on society can be tremendous. Using a "non-linear" model when the true response is "linear" can result in an unacceptable risk to the exposed population. Using a "linear" model when response is "non-linear" could result in banning the use of a beneficial product. Thus, a major question in quantitative risk assessment is how best to describe the shape of the dose-response curve in the low-dose region.

Research in quantitative risk assessment has recently focused on the use of mechanistic information when determining the appropriate tumor incidence model. This work has centered around mechanistic models of carcinogenesis which are characterized as having biologically interpretable parameters. Since some mechanisms are thought to be low-dose linear and some are not, it is expected that knowledge of the carcinogenic mechanism could result in an improved estimate of the shape of the dose-response curve. To facilitate a discussion of these approaches, a four-stage model of carcinogenesis will be used (Anderson, 1987, Bogen, 1989; Portier, 1989; Portier, Hoel, Kaplan and Kopp, 1989).

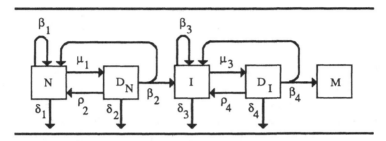

FIGURE 2: A Four-Stage Model of Carcinogenesis With Clonal Expansion
and Repair

Figure 2 shaows a four stage model of carcinogenesis. The model utilizes 5 cell types: normal cells (N), intermediate cells (I), malignant cells (M), damaged normal cells (D_N) and damaged intermediate cells (D_I). The model can be described as follows. Normal cells are allowed to divide and die or differentiate via a birth and death process where β_1 represents the rate of cell division (births) and δ_1 represents the rate at which cells die or differentiate (for brevity, we will refer to this as a removal rate). Normal cells transform into damaged normal cells via some type of genetic aberration (e.g., formation of DNA adducts, single strand breaks, gene amplification, chromosomal translocation) at a rate denoted by μ_1. The genetic aberrations in damaged normal cells are assumed to repairable (e.g. single strand breaks) at the rate ρ_2, returning the cell to its normal state. Damaged cells are also allowed to replicate via a simple birth and death process with rates β_2 and δ_2 respectively. When cell division occurs in an initial damaged cell, the DNA damage is fixed in one of the daughter cells resulting in the creation of a single intermediate cell. The other daughter cell was derived from the strand of DNA without damage and is thus a normal cell.

In intermediate cells, it is assumed that any further DNA damage can no longer be repaired, resulting in an irreversible mutation. It is not necessary for a damaged cell to undergo mitosis in order for a mutation to occur. Many other possible mechanisms such as improper DNA repair can result in permanent alterations to the DNA of a particular cell.

In addition, it is possible that some epigenetic events may lead to permanent alterations in cellular function which might ultimately lead to carcinogenesis. Other models must be proposed before we can study the low-dose behavior of these mechanisms.

Intermediate cells follow a process identical to that of the normal cells where they can divide, be removed or be damaged. The rates of these events are β_3, δ_3 and μ_3 respectively. Damaged intermediate cells can undergo repair (reversion to the intermediate cell state), death/differentiation or birth (resulting in one intermediate cell and one malignant cell) at rates ρ_4, δ_4 and β_4, respectively.

These mathematical models are useful in estimating a number of characteristics of the carcinogenic process, such as the age-specific tumor incidence rate, the expected number of cells of any given type at any given age and/or the distribution of clone sizes for a particular cell type for a given age. The focus of this part of my talk is on estimating the probability of one or more malignant cells by a specified time given a specified exposure history, usually referred to as the distribution of tumor onset times. Mathematical approximations (e.g. Whittemore and Keller, 1978) and numerical tools (e.g. Kopp and Portier, 1989) exist for determining and/or approximating this probability from such models.

In order to use this model effectively for carcinogenic risk estimation, the impact of a potential chemical carcinogen on the various transition rates must be incorporated into the mathematical model. Several mechanisms have been proposed for chemical carcinogenesis that we will consider in the context of this model, emphasizing the implications of the assumed mechanism on the slope of the dose-response curve at dose zero (i.e. low-dose linear mechanisms vs low-dose non-linear mechanisms).

One mechanism by which chemicals may increase cancer risks in animals is by modifying the rate of mitosis of cells in a specific tissue or organ. This modification could be due to numerous factors. The most commonly assumed mechanism is the induction of regenerative hyperplasia in response to tissue damage (e.g. Lewis and Adams, 1987). An increase in the rate of mitosis of this type would represent an increase in the cellular birth rate, β_i, for all stages of cell progression. It follows that increasing the birth rate of initial damaged cells with no subsequent increase in the repair rate will increase the probability of the formation of an intermediate cell. The same phenomenon would occur for final damaged cells, increasing the probability of seeing one or more malignant cells. The relationship between dose and the increase in the birth rates is usually assumed to be non-linear, possibly even having a threshold dose level below which there is no increase in birth rate (e.g. Sakata, et al., 1988). The shape of the dose-response curve in this situation is tied to two events; (i) the relationship between dose of the chemical and the degree of cytotoxicity (e.g. increases in the δ_i) and (ii) the relationship between degree of cytotoxicity and the increase in birth rates. These phenomena need to be studied in more detail before the shape of the resulting dose response curve can be determined.

The transformation from normal cells to intermediate-stage cells can result from damage to cellular DNA which remains unrepaired prior to mitosis (Barrett and Wiseman, 1987). The usual approach to modelling changes in mutation rates from exposure to genotoxic compounds assumes that a single molecule could conceivably "hit" the DNA resulting in damage, a theory which assumes low-dose linearity. However, as noted by Swenberg et al. (1987), the formation of promutagenic damage is only one aspect of a mutation rate. The fixation of the DNA damage is required and is a process related to cell turnover. The cell turnover rate may need to be accelerated in order to fix the DNA damage (e.g., formaldehyde carcinogenesis may act in this manner). The shape of the dose-response relationship between mutations (meaning DNA damage and fixation of this damage) and tumor incidence requires additional investigation.

It is possible that some chemicals may specifically alter the proliferation rate of one specific cell type (e.g. intermediate cells) without altering the proliferation rate of normal cells. Thorslund, Brown and Charnley (1987) contend that chemicals which act on intermediate cells in this manner are mitogens which increase the population of intermediate cells through clonal expansion. Prehn (1964) suggests other possible mechanisms resulting from selective cytotoxicity. These mechanisms, generally referred to as promotion mechanisms, are usually assumed to relate dose to response in a threshold-like manner and thus, dose-related effects on these mechanisms might be modelled using a non-linear change in β_3 as a function of dose. Empirical results supporting a threshold model in favor of a linear model for promotion effects of this type are unavailable and additional work is necessary.

Chemicals may also alter the transformation rate from intermediate cells to malignant cells. Moolgavkar (1983) proposes one mechanism based upon the induction of homologous chromosome exchange during mitosis. Barrett, et al. (1989) suggest that the deactivation of a suppressor gene may be the second mutation in a two-stage carcinogenic process. Very little information exists on how chemicals might alter the rates for this second mutation, thus it is unknown whether chemicals acting in this manner would result in linear or nonlinear dose-response.

The model presented in Figure 2 also allows for a variety of other effects such as the inhibition of DNA repair and/or multiple effects by a single chemical. It is unclear whether chemically induced changes in these rates would be linear or nonlinear in the low dose range. This question is currently under study.

There are numerous statistical problems associated with the use of mechanistic models in quantitative risk assessment for any chemical or chemical mixture. One question concerns whether a chemically induced increase in any of the rates in the model in Figure 2 are independent of the background rate or are adding to the existing rate. The answer to this question can have a serious effect upon the eventual shape of the dose-response curve (Hoel, 1980; Portier, 1987). Some independent effects, even if they are linearly related to dose, can result in low-dose nonlinear behavior, whereas additive linear effects are certain to be low-dose linear. Biological methods are being developed which address the question of whether or not chemically induced carcinoma are somehow different from spontaneously induced carcinoma. For example, Reynolds, et al. (1987) studied the pattern of activated oncogenes occurring in B6C3F$_1$ mouse liver tumors in untreated animals and animals exposed to furan or furfural . Their results suggested that at least part of the increase in liver tumors after exposure to furan or furfural resulted from the induction of weakly activating point mutations in *ras* oncogenes which were not observed in spontaneously occurring tumors. This suggests (at least in part) that the mechanisms by which furan and furfural induce liver tumors are independent of the mechanisms resulting in spontaneous lesions. Independent effects model should be studied for these chemicals.

Another problem concerns the identifiability of carcinogenic mechanism using the most likely source of information, tumor incidence data from the two-year animal carcinogenicity experiment. Using a simple multi-stage model with only two stages where the process of generating a mutation is collapsed into a single step instead of two, Portier (1987) examined the ability of tumor incidence data to accurately differentiate between a chemically induced increase in the rate of mutations from normal cells to initiated cells (this model is "low-dose linear") and a chemically induced increase in the birth rate of initiated cells (a "non-linear model"). It was shown that the probability of incorrectly classifying one mechanism as the other was quite high, approaching 50% in some cases. Portier and Edler (1989) studied the goodness-of-fit of this simple two-stage model to simulated data where the dose effect was (1) linear and on the first mutation step, (2) linear or nonlinear on the birth rate of initiated cells, or (3) linear or quadratic on the second mutation step. For the usual design of the carcinogenesis bioassay, they found that the models (1) through (3) listed above all adequately fit this type of data regardless of the underlying model except when there was a strong nonlinear effect on the birth rate of initiated cells. In this case, linear effects on the two mutation rates could be rejected in up to 40% of the cases. By modifying the design to incorporate start/stop dosing, they were able to improve the model specificity slightly.

The complexity of biologically-based mathematical models may result in the need for an approximation when estimating the tumor incidence rate or some other toxicological endpoint. Several authors (Moolgavkar and Dewanji, 1988; Moolgavkar, Dewanji and Venzon, 1988; Kopp and Portier, 1989) have shown that when these approximations are inappropriate, estimation errors will occur. These estimation errors can result in an incorrect interpretation of the resulting model parameters and/or a faulty prediction of a chemical's toxicity in as-yet-untried experiments.

Figure 2 is no doubt a simplification of what is actually happening in test animals and humans with chemically induced tumors. The methods by which a chemical induces a mutation may vary in different tissues and in different species and may be synergistically related to the presence or absence of other agents in the tissue. It is possible that chemicals which individually yield dose-response relationships that are non-linear in the low-dose region would, in combination, yield a low-dose linear dose-response relationship. The converse is also possible where the individual chemicals have linear dose-response relationships but, in combination are non-linear. Different chemicals may also affect different rates in the multistage process shown in Figure 2 (Pitot, Barsness and Kitagawa, 1978, Farber, 1984). In this case, individual chemicals might be non-carcinogenic, but in combination could prove to be highly carcinogenic (e.g. Schwartz, Pearson, Port and Kunz, 1984). When chemically induced changes in DNA repair are considered, it is also possible that exposure to a single chemical increases carcinogenic risk, but in combination with some other chemical, the risk is reduced.

Finally, any analysis using these mechanistic models is dependent upon the accuracy of the biological theory encompassed by the model. There is no doubt that the model shown in Figure 2 is a simplification of the carcinogenic process. We should always be wary of placing too much emphasis on the mechanistic nature of these models and recognize the limitations of making predictions based upon a curve-fitting exercise.

3. Mechanistically-Based Models For Estimating An Equivalent Dose

Pharmacology, physiology and biochemistry provide the biological theory concerning the estimation of internal dose after external exposure to a potential carcinogen. From these fields, several measures have been proposed as choices for an equivalent dose (ED). The simplest model assumes that tumor response is equivalent on the administered dose scale (D) so that no conversion is necessary. The EPA has sometimes used the average daily dose per unit surface area as the ED for their risk assessments (EPA, 1986). This value is usually calculated by multiplying the administered dose expressed in average daily dose per unit body weight by body weight to the 1/3 power, since the average surface area can be approximated by a formula which is proportional to body weight raised to the 2/3 power (Travis and White, 1988). These models do not have a strong mechanistic basis to them.

When the pharmacology of a compound is known, other measures have been proposed. For example, some chemicals must be transformed to reactive metabolites which then bind to DNA and initiate the carcinogenic process. In these cases, the level of DNA adducts would seem to be a reasonable choice for the ED (Hoel, Kaplan and Anderson, 1983). However, many compounds seem to have both activating and deactivating metabolic pathways. These pathways compete for the available chemical in the tissue. Some metabolic pathways seem to follow linear pharmacokinetics in that the rate at which the metabolite is formed is a linear function of the amount of unmetabolized compound available in the tissue. In this case, the relationship between the administered dose (AD) and the equivalent dose (ED) would also be approximately linear.

Other metabolic reactions can only take effect in the presence of a catalysis and thus follow a Michaelis-Menten-type model (Hoel, Kaplan and Anderson, 1983). In this model, the rate at which the reactive metabolite is formed is linear in dose for small amounts of unmetabolized chemical and becomes constant (or saturates) for higher doses. The resulting relationship between the administered dose of the chemical compound and the reactive metabolite is nonlinear, climbing more rapidly than linear in the low-dose region.

Hoel, Kaplan and Anderson (1983) have shown that when there are competing pathways for the metabolism of a chemical, the relationship between administered dose and effective dose can achieve a variety of shapes. For example, it is possible to get a threshold-looking relationship between the external exposure level and the equivalent dose if there is a very efficient deactivation pathway for a chemical compound which becomes saturated at high doses.

Once a metobolite of a chemical has been identified as the possible toxic agent , it is sometimes possible to model how that metobolite affects normal cell function. Along these lines, one possible ED is the increased rate of mitosis or cell turnover induced by a chemical, usually as a result of chemical mitogenesis (Prehn, 1964). Swenberg, Richardson, Boucheron and Dryoff (1985) proposed using a quantity called the "initiation index", which is the product of the cell replication rate and the level of DNA adduct formation, to allow for both DNA damage and cell turnover.

A problem arises from using pharmacokinetic models in the risk estimation process. Most of the parameters used in these pharmacokinetic models arise from a variety of experiments which must be combined in order to fully characterize the model. This creates a problem for estimation of the variability of carcinogenic risk estimates. To consider all sources of variability in this context is a difficult undertaking. An illustration of one possible approach is given by Portier and Kaplan (1989). In this case, a combination of bootstrapping techniques and Monte Carlo simulation techniques were used to resample from the observed data or to randomly sample from a distribution of possible values for a particular parameter. The example considered by Portier and Kaplan (1989) concerned a physiologically-based pharmacokinetic model for the distribution and metabolism of methylene chloride in mice and humans (Andersen, Clewel, Gargas, Smith and Reitz, 1987). This model is shown in Figure 3.

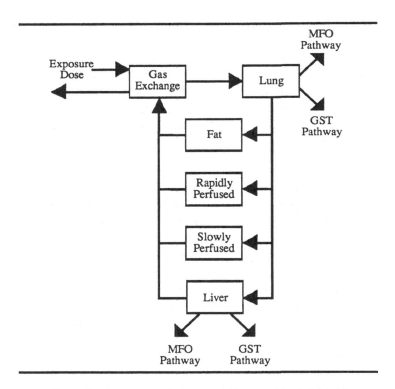

FIGURE 3: The Physiologically-Based Pharmacokinetic Model for Methylene Chloride Proposed by Andersen et al. (1987)

Andersen et al. (1987) proposed that the per-day average concentration of an appropriate metabolite of methylene chloride is a better choice for an ED than the mg/kg/unit-surface-area dose. Estimates of the average concentrations of metabolic products were calculated from the physiologically based pharmacokinetic (PBPK) model illustrated in Figure 3. This PBPK model groups the tissues of the body into five classes which have similar physiology; lung, fat, liver, richly perfused tissues and slowly perfused tissues. The transport of DCM through the various tissues via the circulatory system follows first-order kinetics. The metabolism of DCM occurs through two pathways in both the lung and the liver. One pathway is dependent upon the oxidation of DCM by mixed function oxidases (MFO) and follows Michaelis-Menten kinetics in the experimental dose range. The second pathway for the metabolism of DCM is mediated by the glutathione-S-transferases (GST) system and is assumed to follow first order kinetics in the dose range of interest. The four equivalent dose measures proposed by Andersen et al. are the daily average of the area under the concentration time curve (AUC) for both types of metabolic products (MFO and GST) in both the liver and the lung. These ED's are referred to as RISK1L (MFO pathway in the liver), RISK2L (GST pathway in the liver), RISK1LU (MFO pathway in the lung) and RISK2LU (GST pathway in the lung). The system of differential equations needed to calculate these ED's is given in detail in Andersen et al. (1987).

Portier and Kaplan's results indicate that there could be a substantial increase in the variability of safe-exposure estimates when using mechanistic models with a large number of model parameters. Figure 4 illustrates this result. The line labelled "Bioassay Data Only" represents the frequency distribution of the safe human exposure estimate when the only source of variability considered is the variability of the tumor incidence data from the animal carcinogenicity experiment on methylene chloride. This is contrasted with the other line, labelled "All Sources" where variability is accounted for in all of the data used in the safe exposure estimates, such as body weights, partition coefficients, metabolic constants as well as the tumor incidence data from the animal carcinogenicity experiment. It is clear there is a substantial amount of variability beyond that related to the tumor incidence data since the range of the safe dose estimates increases from less than one order of magnitude to over 3 orders of magnitude. For more details on how these distributions were derived, see Portier and Kaplan (1989).

The increased variability exemplified in Figure 4 is a direct consequence of allowing all (or most) model parameters to vary across individuals. See Harris (this volume) for a discussion of the interindividual variability in activation and inactivation pathways and in DNA repair capacity. This increased variability is not a shortcoming of the use of these more complicated models. On the contrary, these detailed mechanistic models may provide insight into the process of carcinogenesis and the necessary pieces of this process. The increased variability observed by Portier and Kaplan may represent a more reasonable estimate of the population variability with respect to the safe exposure level. This variability should be included in order to estimate not only the mean safe exposure level, but the entire distribution of safe exposures in the human population.

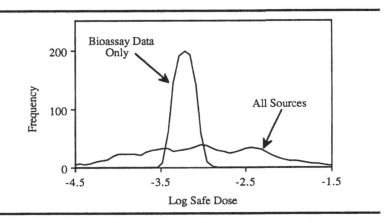

FIGURE 4: An Illustration of the Impact of Multiple Sources of Variability on the Overall Distribution of the Estimated Safe Exposure Level Derived By Portier and Kaplan (1989)

4. Summary

This paper has described the basic format used in estimating risks from exposure to a chemical carcinogen using toxicology data from animal and human studies. One class of mathematical models used in this context was described in detail; the multistage model of tumor incidence with clonal expansion of cell populations. There are numerous other models used in the risk estimation process which include other tumor incidence models (e.g. the gamma multi-hit model), models relating administered dose to the equivalent dose (e.g. physiologically-based pharmacokinetic models), and allometric formulae relating simple body measurements (e.g. body weight) to items important to chemical carcinogenesis (e.g. liver weight). In all cases, the current research emphasis is on models which presumably have a higher degree of biological plausibility than those used previously. Because of this emphasis, these models will also possess many of the problems discussed for the multistage model in Figure 2; most notably, a lack of identifiability of model parameters, an inability to determine linearity versus non-linearity in the low-dose region and difficulty in assessing the overall variability of risk estimates derived from multiple experiments. When chemical mixtures and multiple exposures are considered, the process of estimating carcinogenic risks will be even more complicated.

This discussion has focused on limitations to the use of mechanistic models in carcinogenic risk assessment. However, cellular biology and biochemistry hold some promise for improving our ability to accurately parametrize mechanistic mathematical models. By incorporating information on the size distribution and number of malignancies and pre-malignant lesions into these mechanistic models, we may be able to improve the differentiation between chemicals which behave in a "low-dose linear" fashion and those which don't. In addition, by radio-labelling cells, we are able to estimate the rate of mitosis in different cell populations; information which may also prove useful in parameterizing mechanistic models. This work is still in a very early stage of development and further work needs to be done before it will be useful for risk assessment.

5. References

Andersen, M., Clewell, H., Gargas, F., Smith, F., and Reitz, R. (1987). Physiologically based pharmacokinetics and the risk assessment process for methylene chloride. Toxicology and Applied Pharmacology 87: 185-202.

Anderson, E. and the Carcinogen Assessment Group (1983). Quantitative approaches in use to assess cancer risk. Risk Analysis 3: 277-295.

Anderson, M. (1987). Issues in biochemical applications to risk assessment: How do we evaluate individual components of multistage models? Environmental Health Perspectives 76, 175-180.

Armitage, P. and Doll, R. (1954). The age distribution of cancer and a multistage theory of cancer. British Journal of Cancer 8: 1-12.

Barrett, J. C. and Wiseman, R. (1987). Cellular and molecular mechanisms of multistep carcinogenesis: Relevance to carcinogen risk assessment. Environmental Health Perspectives 76, 65-70.

Barrett, J. C. and Wiseman, R. W. (1989). Relevance of Cellulart and Molecular Mechanisms of Multistep Carcinogenesis to Risk Assessment. (unpublished manuscript).

Bogen, K. (1989). Cell proliferation kinetics and multistage cancer risk models. Journal of the National Cancer Institute 81, 267-277

EPA (1986). Guidelines for carcinogen risk assessment. 51 Federal Register 33992, 1-17.

Farber, E. (1984). Cellular biochemistry of the stepwise development of cancer with chemicals. Cancer Research 44: 5463-5474.

Hoel, D., Kaplan, N. and Anderson, M. (1983). Implication of nonlinear kinetics on risk estimation in carcinogenesis. Science 219: 1032-1037.

Hoel, D. (1980). Incorporation of background in dose-response models. Federation Proceedings 39: 73-75.

Kopp, A. and Portier, C. (1989). A note on approximating the cumulative distribution function of the time to tumor onset in multistage models. (unpublished manuscript).

Lewis, J. and Adams, D. (1987). Inflammation, oxidative DNA damage and carcinogenesis. Environmental Health Perspectives 76, 19-28.

Moolgavkar, S. (1983). Model for human carcinogenesis: Action of environmental agents. Environmental Health Perspectives 50: 285-291.

Moolgavkar, S., Dewanji, A., and Venzon, D. (1988). A stochastic two-stage model for cancer risk assessment. I. The hazard function and the probability of tumor. Risk Analysis 8 (3), 383-392.

Moolgavkar, S. and Dewanji, A. (1988). Biologically based models for cancer risk assessment: A cautionary note. Risk Analysis 8 (1), 5-6.

Pitot, H., Barsness, L. and Kitagawa, T. (1978). Stages in the process of hepatocarcinogenesis in rat liver. Carcinogenesis 2: 433-442.

Portier, C. (1987). Statistical properties of a two-stage model of carcinogenesis. Environmental Health Perspectives 76: 125-131.

Portier, C. and Kaplan, N. (1989). The variability of safe dose estimates when using complicated models of the carcinogenic process. A case study: Methylene chloride. Fundamental and Applied Toxicology 13, 533-544.

Portier, C. (1989). Quantitative risk assessment. In Ragsdale, N. and Menzer, R. (eds.): **Carcinogenicity and Pesticides**. American Chemical Society Symposium Series Number 414; (to appear).

Portier, C., Hoel, D., Kaplan, N. and Kopp, A. (1990) Biologically based models for risk assessment. In H. Vannio, M. Sorsa and A. J. McMichael (eds) **Complex Mixtures and Cancer Risk**. IARC Scientific Publications Number 104, International Agency for Research on Cancer, Lyon (to appear).

Portier, C. and Edler, L. (1989) Two-stage models of carcinogenesis, classification of agents and design of experiments. (submitted)

Prehn, R. (1964). A clonal selection theory of chemical carcinogenesis. Journal of the National Cancer Institute 32 (1): 1-17.

Reynolds, S. Stowers, S., Patterson, R., Maronpot, R., Aaronson, S. and Anderson, M. (1987). Activated oncogenes in B6C3F1 mouse liver tumors: Implications for risk assessment. Science 237, 1309-1316.

Sakata, T., Masui, T., St. John, M. and Cohen, S. (1988). Uracil-induced calculi and proliferative lesions of the mouse urinary bladder. Carcinogenesis 9 (7), 1271-1276.

Schwartz, M., Pearson, D., Port, R. and Kunz, W. (1984). Promoting effect of 4-dimethylaminoazobenzene on enzyme altered foci in rat liver by N-nitrosodiethanolamine. Carcinogenesis 5: 725-730.

Swenberg, J., Richardson, F., Boucheron, J. and Dryoff, M. (1985). Relationships between DNA adduct formation and carcinogenesis. Environmental Health Perspectives 62: 177-183.

Swenberg, J., Richardson, F., Boucheron, J., Deal, F., Belinsky, S., Charbonneau, M. and Short, B. (1987). High to low dose extrapolation: Critical determinants involved in the dose response of carcinogenic substances. Environmental Health Perspectives 76, 57-63.

Thorslund, T., Brown, C. and Charnley, G. (1987). Biologically motivated cancer risk models. Risk Analysis 7: 109-119.

Travis, C. and White, R. (1988). Interspecies scaling of toxicity data. Risk Analysis 8, 119-125.

Whittemore, A. and Keller, J. (1978). Quantitative theories of carcinogenesis. SIAM Review 20, 1-30.

Variability of Unit Risk Estimates under Different Statistical Models and between Different Epidemiological Data Sets

Heiko Becher [1] and Jürgen Wahrendorf [1]

Abstract

The unit risk has been proposed as a conceptually simple parameter which allows to describe the effect of a carcinogenic substance in low concentration in ambient air to the general population. It is in frequent use by agencies like the WHO or the EPA. In this paper we describe some models for dose-response analysis and methods for estimating the unit risk and investigate the variability of unit risk estimates as it emerges from using different dose-response analyses and various datasets. Three substances are investigated, two of which were found to provide a sufficient basis for quantitative risk estimation (arsenic, benzene). The systematic variation of the unit risk estimates was found to be relatively small for both substances.

[1]   Institute of Epidemiology and Biometry, German Cancer Research Center, D-6900 Heidelberg, Federal Republic of Germany

1.   **Introduction**

Quantitative risk estimation although not always loved by scientists has become an indispensible component of regulatory approaches to cancer prevention. Following a qualitative evaluation of the carcinogenic risk of a given exposure to humans which may involve data from short-term tests, animal studies and epidemiological investigations various ways of quantitative risk estimation can be followed. In general, these include thorough dose-response analyses of experimental and epidemiological studies. These analyses are often combined with population data on morbidity or mortality and with data on environmental exposure. For experimental studies more precise exposure information is usually at hand, but the assumptions to be made to transfer the results to the human situation are enormous. Epidemiological studies have the advantage of being immediately applicable to the human situation. Exposure levels measured in specific studies are not too far away from the range in which regulatory limits have to be established.

Dose-response analyses of epidemiological data can be based on the assumption of various effect measures (absolute risk, relative risk) and statistical models (additive, multiplicative and others). Quantitative risk assessment can only be performed if a single easily interpretable estimate results from the excercise. The "unit risk" has been considered for this purpose but has been frequently critizised for the potential that the estimates may vary considerable with varying specification of the statistical model, other assumptions or data sets.

The estimation of relative risks has become a well established concept and method for the analysis of epidemiological data [5,6]. The derivation of single parameters which quantify the dose-response relationship and can be used for risk assessment needs still to be explored. In this paper we want to address the question whether the effect a substance has on the mortality/morbidity of a population can usefully be summarized through a single parameter. The unit risk (UR) is considered for this purpose. However, before unit risk estimation several other assumptions are already made, some of which are of statistical nature, others address more general questions, such as the selection of appropriate data. These steps which are a necessary part in quantitative risk estimation will be discussed in some detail. Quantitative risk estimation using the unit risk parameter will be illustrated for three different compounds (arsenic, benzene, nickel).

2. Dose-response-Analyses

There are a number of textbooks in which issues of analysing epidemiological data [5,6] or experimental data [13] are described in detail. In this paper we briefly describe some models which turn out to be essential for the compounds considered here.

Data which are relevant for arsenic and benzene risk assessment stem mainly from occupational cohort studies. From such studies the following data are normally available:

- date of birth
- occupational history (duration in certain jobs, level(s) of exposure)
- end of follow-up (data and cause of death, termination of study lost to follow-up)
- other covariates (sex, smoking etc.)

External data (mortality rates) are often used to calculate the expected number of deaths. If the original data contain information on the exposure levels during the time at work, various possibilities exist to define an appropriate dose variable and to incorporate this into a dose-response analysis. The following methods are often used:

- cumulative exposure
- time weighted cumulative exposure
- cumulative exposure with a lag period
- maximum exposure
- average exposure

If biological knowledge does not speak in favour of a particular method, the choice of the method used is usually based on statistical arguments like goodness-of-fit statistics.

For case-control data similar considerations hold. However, those data are normally less detailed with respect to the exposure of interest.

Modern dose-response analyses are based on various regression techniques. As an example we outline a method which is applicable for grouped cohort data, the so-called Poisson regression. For ease of presentation we assume that there are J age groups and K exposure categories. Let λ_{jk} be the mortality rate of the j-th age group and the k-th exposure category, a_j a nuisance parameter for the effect of the j-th age group, x_k the exposure level for the k-th exposure group and β the parameter of interest which describes the effect of the exposure. A poisson regression model is then given as:

$$\log (\lambda_{jk}) = a_j + \beta x_k$$

or

$$\log (\lambda_{jk}) = \log (\lambda_j^*) + \log(\mu) + \beta x_k$$

In the second model, λ_j^* is the mortality rate in the general population for age group j and μ is the ratio of the mortality rate for unexposed persons in the study population and the mortality rate in the general population. It follows from these models that the relative risk for exposure group k compared to exposure group k' is given by

$$\frac{\lambda_{jk}}{\lambda_{jk'}} = \exp\left[\beta\left(x_k - x_{k'}\right)\right]$$

Thus, relative risk functions concerning an exposure x as compared to unexposed are of the form

$$RR\ (x) = \exp\ (\beta x).$$

When fitting these models using software packages like GLIM or EGRET, the observed number of death in a cell is assumed to be a poisson-variable and the logarithm of the person-years or the expected number of deaths (in the first and second model respectively) offsets the regression equation.

If an analysis is to be performed on the basis of individual data, an extension of the Cox-Model [8] may be used for the analysis. Further details of the Cox-regression model for cohort studies may be found in [6].

For case-control data logistic regression models are the common class of models which allow dose-response analyses provided the data are of sufficient detail. In this case, various relative risk functions may be used:

$$RR(x) = \exp(\beta\ x) \qquad\qquad \text{log-linear Model [8]}$$
$$RR(x) = 1 + \beta\ x \qquad\qquad \text{linear Model [3]}$$

$$\ln RR\ (x) = \begin{cases} \left[\left(1 + a\left(\beta\ x\right)\right)\right]^{1/a} & a \neq 0 \\ \exp\left(\beta\ x\right) & a = 0 \end{cases} \qquad \text{mixture model [14]}$$

where x is the variable (or a vector containing the variable) which describes the exposure, for instance the cumulative exposure. It is also possible to use a function of the exposure variable (e.g. square root).

3. Unit risk - definition and estimation

The basis definition of the "unit risk" (UR) is fairly simple. It is the additional (above background) probability of developing a disease (or dying of a disease) under a life-long exposure to 1 $\mu g/m^3$ of the substance of interest.

$$UR = Pr(D|\text{lifelong exposure to } 1\mu g/m^3) - Pr(D|\text{no exposure})$$
$$= P_1 - P_0$$

In the literature different methods to estimate the unit risk are suggested. For all methods the baseline incidence or mortality rates for the disease of interest and for all causes of death are required. Usually the mortality rates are given per 100 000 for seventeen 5-year intervals up to the age of 85 and one final open age-interval (85+). Then P_0 may be estimated as

$$P_0 = \sum_{i=1}^{17} 5\, r_i / 100000 \; \pi \prod_{j=1}^{i-1} (1 - 5\, s_j / 100000)$$

$$+ \; r_{18} / s_{18} \; \pi \prod_{j=1}^{17} (1 - 5\, s_j / 100000)$$

where r_i is the mortality rate of the disease of interest in the age interval i and s_i is the overall mortality rate in the age interval i. In the Federal Republic of Germany, P_0 for lung cancer is 0.049 for males and 0.009 for females which gives an average value for the total population of 0.029 [2]. The corresponding average value for the USA is $P_0 = 0.0451$ [11].

The WHO [23] uses a very simple method to estimate the unit risk. It assumes that the relative risk is a linear function of the cumulative dose and it is applicable if the relative risk of the study population compared to the unexposed and the mean cumulative exposure is the only available information. Then

$$UR_{WHO} = P_0 \left(1 + \frac{RR-1}{X}\right) - P_0 = P_0 \frac{RR-1}{X}$$

where P_0 is defined as above, RR is the estimated relative risk and X is the lifetime exposure (standardized lifetime exposure for the study population on a lifetime continuous exposure basis): in the case of occupational studies, X represents a conversion from the occupational 8-hour, 240-day exposure over a specific number of working years and can be calculated as $X = TWA \cdot 8/24 \cdot 240/365 \cdot$ (average exposure duration [in years]) / (life expectancy [70 years]), where TWA is the 8-hours time-weighted average ($\mu g/m^3$), or, equivalently, X = cumulative exposure ($\mu g/m^3 \cdot$ years) $\cdot$ 8/24 $\cdot$ 240/365 / (life expectancy [70 years]).

This method has the advantage that simple published results can often be used directly. Dose-response analyses as outlined in the previous section are not required because a linearity is assumed. There are, however, certain limitations of the method. The linearity of the dose-response curve may not be a sufficient approximation of the true relationship, there are indeed some studies which speak in favour of a concave downward, concave upward or even non-monotonous dose-response function. If in a publication the relative risk estimation of different exposure subgroups are given, this information is not used with the WHO-method.

In a more sophisticated method for estimating P_1 the mortality rates are combined with relative or absolute risk estimates derived under a chosen statistical model. This can be developed with individual-based original data from epidemiological studies as well as with published aggregate estimates of relative risks.

A conversion from occupational exposure to environmental exposure is also required. A concentration of k $\mu g/m^3$ is the occupational setting is assumed to be equivalent to a concentration of k $\cdot$ 8/24 $\cdot$ 240/365 $\approx$ k $\cdot$ 0.22 in the environmental air.

Suppose the dose-response function RR(x) is derived from the data using one of the previous described methods. Under the assumption that the relative risk relationship holds over all age groups, then $r_i(x)$, the mortality rate for age group i for the disease of interest under an exposure x, is given by

$$r_i(x) = r_i \cdot RR(x)$$

Similarly, if an excess risk model is assumed, then

$$r_i(x) = r_i + ER(x)$$

and $s_i(x)$, the mortality rate for all causes for age group i is given by

$$s_i(x) = s_i + r_i(x) - r_i$$

The probability for the disease of interest given a (cumulative) exposure x may then be estimated as

$$P_X = \sum_{i=1}^{17} 5\, r_i(x) / 100000 \; \prod_{j=1}^{i-1} (1 - 5\, s_j(x) / 100000)$$

$$+ \, r_{18}(x) / s_{18}(x) \; \prod_{j=1}^{17} (1 - 5\, s_j(x) / 100000)$$

It has to be noted that x varies over time intervals because the cumulative exposure increases with increasing age. This is indicated by a subscript i.

If we assume a 5-year lag period and a constant exposure of k $\mu g/m^3$ then the effective cumulative exposure x_i for age interval i is approximately x_i = $k \cdot 5(i-1)\mu g/m^3 \cdot$ years. For unit risk estimation the effective cumulative exposure x_i for age interval i assumes a constant exposure of $1\mu g/m^3$ and is thus given by $x_i = 5(i-1)$ $\mu g/m^3 \cdot$ years and the probability for the disease under this exposure may be estimated using x_i accordingly in the above equation. The unit risk estimate is then the difference of both estimated probabilities,

$$UR = P_1 - P_0.$$

In some cases the data available do not allow a detailed dose-response analysis according to the methods in section two, however, relative risks are given for at least two exposure subgroups. In order to take advantage of this information, the following estimate for the UR is proposed. We assume that for each exposure subgroup i the mean exposure, the relative risk RR_i and the person-years (PY) are given. Then

$$UR = P_0 \cdot \sum_{i=1}^{k} \frac{RR_i - 1}{X_i} \frac{PY_i}{PY}$$

This is a weighted mean of the unit risk estimates from each subgroup using the WHO-method and we denote this method by 'extended WHO-Method'. The weights are given according to the number of person-years in each group. Alternatively, if the person-years are not given in the publication, the expected number of deaths may be used. The difference can be assumed to be negligible.

This method also allows simple check of the assumption of the linearity of the relative risk function. The unit risk estimates derived from each subgroup should be the same or at least in the same order of magnitude. Remarkable systematic deviations from this may be indications of a non-linear dose response relationship.

4.   Arsenic

Arsenic is well-known as a carcinogen for man [16].   Epidemiological studies on arsenic-exposed occupational cohorts clearly showed an excess of lung cancer cases, see, for example, [9,10,17,19].   In some cohort studies information on exposure levels are available and were therefore considered appropriate for quantitative risk assessment by agencies such as EPA [19] or WHO [23].

For the purpose of this excercise we used data from four major cohort studies characteristics of which are summarized in Table 1 with their most recent references.

TABLE 1:   Cohort Studies on Arsenic Exposure

study population	basic characteristics	recent reference
Anaconda Copper Smelter	Follow-up period 1938-1977, 8047 white males, 302 respiratory cancer deaths, 106 expected (SMR=285, p<0.01), industrial hygiene data on arsenic concentration collected between 1943 and 1958	[17]
Tacoma Copper Smelter	Follow-up period 1941-1977, 2802 males, 104 respiratory cancer deaths, 39.6 expected (SMR=262.6, p<0.01), measurements since 1938 and urinary measurements since 1948	[9]
Insecticide producing Company	Follow-up period 1940-1973, 603 males, 20 respiratory deaths, 5.8 expected (SMR=343, p<0.01) air samples taken in 1943 and in 1952	[19]
Eight Copper Smelters	Follow-up period 1949-1980, 6078 white males, 93 lung cancer deaths, air arsenic levels available from 6 smelters	[10]

Individual-based data of the Anaconda study were kindly made available to us by Dr. J. Lubin of the NCI. In all cases published aggregate estimates of relative risk were also used for reanalysis. Lung cancer mortality data for the Federal Republic of Germany [2] were used for unit risk estimation.

Different dose response models relating the information on exposure to the relative risk of dying of lung cancer were fitted to the data from the four cohorts. The results are summarized in Table 2. Linear, exponential or power function models were fitted either to individual-based data, published aggregate data or derived directly from publications as indicated. Unit risk estimates were then derived according to the procedures described above. Frequently modifications of the model were made by different categorizations of dose as well as applying different inclusion criteria.

TABLE 2: Unit Risk Estimates for Arsenic

reference	model	dose-response-function		unit risk estimate
Anaconda Copper Smelter				
[18]	L	$1+0.0001175\,x$	(b)	0.00097
	L	$1+0.0002003\,x$	(b)	0.0017
	L	$1+0.0003389\,x$	(b)	0.0028
[24]	L	$1+0.0003173\,x$		0.0026
	E	$\exp(0.0001142\,x)$		0.00096
[7]	E	$\exp(0.00038\,x)$	(*)	0.00032
[17]	L	$1+0.0003754\,x$		0.0031
	E	$\exp(0.0002748\,x)$		0.0024
[9]	P	$1+0.04062\,x^{0.3843}$	(*)	0.0098
author's	E	$\exp(0.00004874\,x)$	(i)	0.00041
analyses	E	$\exp(0.0003416\,x^{1/2})$	(i)	0.0017
	L	$1+0.0003228\,x$	(b)	0.0027

Tacoma Copper Smelter

[9]	P	$1+0.04897 \, x^{0.3499}$	(*)	0.0097
	P	$1+0.04857 \, x^{0.4081}$	(*)	0.013
	L	$1+0.0007359 \, x$		0.0068

8 Different Copper Smelters

[10]	L	$1+0.0005237 \, x$	(b)	0.0043
	L	$1+0.0003812 \, x$	(b)	0.0031
	E	$\exp(0.001 \, x)$	(*)	0.0095
	E	$\exp(0.00081 \, x)$	(*)	0.0075
	E	$\exp(0.0439 \, x^{1/2})$	(*)	0.030
	E	$\exp(0.0093 \, x^{1/2})$	(*)	0.0048

Michigan Pesticide Factory

[19]	L	$1+0.0003629 \, x$	(b)	0.0030
	L	$1+0.0009224 \, x$	(b)	0.0075

x-cumulative dose in $\mu g/m^3$ years
L-linear relative risk model (b)-different measures of dose
E-exponential relative risk model (i)-individual based data
P-power function as relative risk model (*)-taken from literature

The resulting 23 estimates of the unit risk are displayed on a logarithmic scale in Figure 1. It can be noted that the majority of estimates falls within one order of magnitude (10^{-3} to 10^{-2}).

FIGURE 1: Unit Risk Estimates for Arsenic

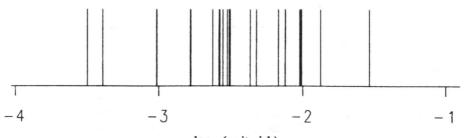

$$\log_{10}(\text{unit risk})$$

The median estimates for the four cohorts are: Anaconda: 0.0021; Tacoma: 0.0097; Eight Smelters: 0.0062; Pesticide Factory: 0.0052. The mean of these medians is 0.0057 which is slightly larger than the overall median or mean. The estimate used by EPA is 0.0043 and we conclude that estimates derived under different statistical models from different epidemiological data sets do not deviate essentially from this estimate. However, it has clearly to be noted that we have only dealt with possible systematic variation of the estimates leaving aside their statistical variation which is composed of variation in exposure assessment as well as variation in the assessment of the mortality endpoint.

5.   Benzene

A leukemogenic effect of benzene has been suspected already since 1928 when case reports have been published. Several epidemiological studies have supported this and lead to the categorization of benzene as a human carcinogen [4]. A review of risk assessments has been given in [1]. For the purpose of quantitative risk estimation we considered four cohort studies. Their main characteristics are summarized in Table 3.

TABLE 3: Cohort Studies On Benzene Exposure

study population	basic characteristics	recent reference
Pliofilm Cohort	Follow-up period 1950-1981, 1162 exposed individuals, 31612 person-years, 9 leukemia deaths observed, 2.66 expected (SMR=337). Industrial hygiene measurements on benzene concentrations	[22]
Dow-Chemical Cohort	Follow-up period 1938-1982, 956 individuals, 4 leukemia deaths observed, 2.1 expected (SMR=194). (Detailed exposure measurements)	[4]

Chemical Industry (different companies) U S A	7676 individuals. Internal non-exposed comparison group available. In exposed group 6 leukemia deaths, 4.44 expected. In non-exposed group no leukemia deaths, 3.4 expected.	[25]
Benzene Workers in China	Follow-up period 1972-1981, 28460 exposed workers (30 leukemia cases), 28257 control workers (5 leukemia cases). Benzene concentrations determined by means of grab samples + gaschromatographic analyses	[26]

Similar to the previous section Table 4 gives the unit risk estimate which were derived from the published data. In only one cohort several statistical models and methods of unit risk estimation could be investigated. Estimates derived under exponential or linear model differ again by not more than one order of magnitude.

The 10 estimates are displayed on a logarithmic scale in Figure 2. Their median is $9.17 \cdot 10^{-6}$. The value agrees well with the "best-judgement" unit risk of $8.1 \cdot 10^{-6}$ derived by the EPA Carcinogen Assessment Group [12].

FIGURE 2: Unit Risk Estimates for Benzene

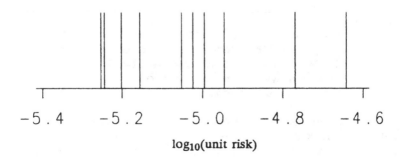

$$-5.4 \quad -5.2 \quad -5.0 \quad -4.8 \quad -4.6$$

$\log_{10}$(unit risk)

TABLE 4: Unit Risk Estimates For Benzene

reference	model	dose-response-function	unit risk estimate
Pliofilm Cohort			
[22]	E	$\exp(3.96 \cdot 10^6\,x)$ (*)	$5.7 \cdot 10^{-6}$
	E	$\exp(5.3 \cdot 10^6\,x)$ (*)	$7.0 \cdot 10^{-6}$
	E	$\exp(4.8 \cdot 10^{-5}\,x^{0.3})$	$9.46 \cdot 10^{-6}$
	L	- (WHO-method)	$2.28 \cdot 10^{-5}$
	L	- (WHO-method)	$1.70 \cdot 10^{-5}$
	L	- (extended WHO-method)	$5.58 \cdot 10^{-6}$
	L	- (extended WHO-method)	$8.88 \cdot 10^{-6}$
Dow-Chemical Cohort			
[4]	L	- (WHO-method)	$1.13 \cdot 10^{-5}$
Chemical Industry Cohort			
[25]	L	- (WHO-method)	$6.29 \cdot 10^{-6}$
China Cohort			
[26]	L	- (WHO-method)	$9.46 \cdot 10^{-6}$

x-cumulative dose in $\mu g/m^3$
L-linear relative risk model   (*)-taken from literature
E-exponential relative risk model

6.   Nickel

Many epidemiological studies demonstrate excess incidences of cancers of the nasal cavity and lung among workers in nickel refineries in different countries. Quantitative risk estimation, however, remains difficult for the following reasons. Many epidemiological studies, in particular those concerned with cohorts exposed some decades ago, lack quantitative information on exposure. Duration of exposure has been used as a surrogate in most analyses. More recent studies are frequently too small to overcome problems in relation to the healthy worker effect. In addition, quantitative estimates of exposure derived from current workplace investigations suggest what a variety of nickel compounds is present in different workplace. Finally, experimental studies in animals demonstrate that the carcinogenic effect can be very different between nickel compounds and lead to the hypothesis that this may be related to the solubility of the compound [20,21].

In essence, the data base for quantitative risk estimation was felt to be too sparse to derive estimates which can be defensible.

7.   Discussion

Quantitative risk estimation is an essential component of risk assessment which leads to setting exposure levels to populations exposed to carcinogens in the workplace or in the general environment.

Many simplifying pragmatic assumptions have to be entered into such a process. It has frequently been questioned whether such simplifications may lead to estimates the variability of which may be very large.

We have investigated how unit risk estimates can differ under different statistical models and between different epidemiological data sets. Data on arsenic, benzene and nickel were reviewed for this purpose. While for nickel it was felt that no reliable estimates could be derived, mainly due to lacking specific exposure information in the studies, we generated a range of some ten or twenty estimates for benzene and arsenic. Obviously, the choice of statistical models or procedure is arbitrary. However, we feel that our range of models has covered those which can reasonably be fitted to the data fairly comprehensively. As far as data sets are concerned all major epidemiological studies were used.

The main conclusion for arsenic and benzene is that the systematic variation of unit risk estimates as investigated here does not go much beyond one order of magnitude. Use of such estimates may thus be considered for risk assessment with some confidence.

However, it has clearly to be noted that statistical variation coming from variability of exposure assessment and endpoint determination may well add to the systematic variation a considerable amount.

In any case, the use of epidemiological data which contain such sources of variation is preferable to animal data where additional assumptions on inter-species conversions need to be made for which again no sound knowledge about the variability is at hand.

Acknowledgement

This work was supported by the Umweltbundesamt (UBA), Berlin (West). Secretarial assistance by Mrs Heike Weis is gratefully acknowledged.

Literatnre:

1. Austin, H., Delzell, E., Cole, P. (1988) Benzene and Leukemia: A review of the literature and a risk assessment. *Am.J.Epidemiol.*, **127**, 419-439.

2. Becker, N., Frentzel-Beyme, R., Wagner, G. (1984) Atlas of Cancer Mortality in the Federal Republic of Germany. Springer, Heidelberg.

3. Berry, G. (1980) Dose-response in case-control studies. *J.Epidemiol.Community Health*, **34**, 217-222.

4. Bond, G.G., McLaren, E.A., Baldwin, C.L., Cook, R.R. (1986) An update of mortality among chemical workers exposed to benzene. *Br.J.Ind.Med.*, **43**, 685-91.

5. Breslow, N.E. & Day, N.E. (1980) Statistical methods in cancer research Vol. I - The analysis of case-control studies. *IARC Scientific Publications No. 32.*

6.   Breslow, N.E. & Day, N.E. (1987) Statistical methods in cancer research Vol. II - The design and analysis of cohort studies. *IARC Scientific Publications No. 82.*

7.   Breslow, N.E., Lubin, J.H., Marek, P., Langholz, B. (1983) Multiplicative models and cohort analysis. *J.Am.Stat.Assoc.*, **78**, 1-12.

8.   Cox, D.R. (1972) Regression models and life tables (with discussion). *J.R.Stat.Soc. B*, **34**, 187-220.

9.   Enterline, P.E., Henderson, V.L. & Marsh, G.M. (1987) Exposure to arsenic and respiratory cancer. A Reanalysis. *Am.J. Epidemiol.*, **125**, 929-938.

10.   Enterline, P.E., Marsh, G.M., Esmen, N.A., Henderson, V.L., Callahan, C.M. Paik, M. (1987) Some effects of cigarette smoking, arsenic, and SO_2 on mortality among US copper smelter workers. *J.Occup.Med.*, **29**, 831-838.

11.   Environmental Protection Agency (1984) The carcinogen assessment group's final risk assessment on arsenic. EPA-600/8-83-02IF.

12.   Environmental Protection Agency (1985) Carcinogen Assessment Group. Interim quantitative cancer Unit Risk estimates due to inhalation of benzene. EPA-600/X-85-022.

13.   Gart, J.J., Krewski, D., Lee, P.N., Tarone, R.E., Wahrendorf, J. (1986) Statistical methods in cancer research Vol. III - The design and analysis of long-term animal experiments. *IARC Scientific Publications No. 79.*

14.   Guerrero, V.M., Johnson, R.A. (1982) Use of the Box-Cox transformation with binary response models. *Biometrika*, **69**, 309-314.

15.   International Agency for Research on Cancer (1982) Evaluation of the carcinogenic risk of chemicals to humans. *IARC Monograph*, 29.

16.   International Agency for Research on Cancer (1987) Overall Evaluations of Carcinogenicity: An Updating of IARC Monographs Volumes 1 to 42. *IARC Monograph*, Suppl. 7.

17. Lee-Feldstein, A. (1986) Cumulative exposure to arsenic and its relationship to respiratory cancer among copper smelter employees. *J.Occup.Med.*, **28**, 296-302.

18. Lubin, J.H., Pottern, L.M., Blot, W.J., Tokudome, S:, Stone, B.J., Fraumeni, J.F. (1981) Respiratory cancer among copper smelter workers: recent mortality statistics. *J.Occup.Med.*, **23**, 779-784.

19. Ott, M.G., Holder, B.B. & Gordon, M.D. (1974) Respiratory cancer and occupational exposure to arsenicals. *Arch. Environ.Health*, **29**, 250-255.

20. Pott, F., Rippe, R.M., Roller, M., Csicsaky, M., Rosenbruch, M., Huth, F. (1988) Carcinogenicity of nickel compounds and nickel alloys in rats by I.P. Injection. Proceed. of the 4th Intern. Conference on Nickel Metabol. and Toxicol., Finland.

21. Pott, F., Ziem, U., Reiffer, F.J., Huth, F., Ernst, H., Mohr, U. (1987) Carcinogenicity studies on fibres, metal compounds, and some other dusts in rats. *Exp.Pathol.*, **32**, 129-152.

22. Rinsky, R.A., Smith, A.B., Hornung, R., Filloon, T.G., Young, R.J., Okun, A.H., Landrigan, P.J. (1987) Benzene and leukemia - an epidemiologic risk assessment. *N.Engl.J.Med.*, **316**, 1044-1050.

23. WHO (1987) Air quality guidelines for Europe. WHO regional publications. European series No. 23.

24. Welch, K., Higgins, I., Oh, M., Burchfiel, C. (1982) Arsenic exposure, smoking, and respiratory cancer in copper smelter workers. *Arch.Environ.Health*, **37**, 325- 335.

25. Wong, O. (1987) An industry wide mortality study of chemical workers occupationally exposed to benzene. II. Dose response analyses. *Br.J.Ind.Med.*, **44**, 382-95.

26. Yin, S.N., Li, G.L., Tain, F.D., Fu, Z.I., Jin, C., Chen, Y.J., Luo. S.J., Ye, P.Z., Zhang, J.Z., Wang, G.C. (1987) Leukemia in benzene workers: A retrospective cohort study. *Br.J.Ind.Med.*, **44**, 124-8.

Carcinogenic Drugs:
A Model Data-Base for Human Risk Quantification

J. Kaldor

ABSTRACT

   Many of the drugs used in cancer chemotherapy are
themselves carcinogenic. However, they differ from other
carcinogens, in that humans are intentionally exposed to
them at high, carefully measured doses. Studies of second
cancers following chemotherapy have as their primary goal
the reduction of long-term risk through suitable
modification of therapy. However, they can also provide
unique quantitative information on human carcinogenesis.
We discuss the information available on various aspects of
chemotherapy-induced leukemia, including dose-effect and
temporal relationships, and suggest ways in which
statistical analyses could be extended. Further topics
considered include the comparative carcinogenicity of
chemotherapeutic agents and radiation, and other endpoints
related to carcinogenicity which may usefully be studied in
treated patients.

1.   INTRODUCTION

   Quantitative estimation of human cancer risk requires
either detailed knowledge of the mechanisms of
carcinogenesis, or a substantial amount of information on
the empirical correlation between risk estimates from
epidemiology and the corresponding risk predictions from
experimental studies. Research into carcinogenic
mechanisms has certainly made substantial progress,
particularly with the recent application of molecular
biological tools. However, we are still very far from
being able to calculate the human cancer risk presented by
an agent simply on the basis of its physical, chemical or
biochemical properties.

   The quantitative predictive value of experimental
studies can only be evaluated if both epidemiological risk
estimates and experimental predictions are available for a
set of agents large enough to allow the study of
correlation between the two. In a recent paper, Allen et
al. [1] identified 23 substances which are carcinogenic to
animals, humans or both, and studied the correlation

between carcinogenic potency estimates derived from epidemiological and experimental studies. The correlation was impressively high, and demonstrated that, within broad bands of uncertainty, animal carcinogenicity results can be used to quantitatively predict human cancer risk. However, the group of agents studied was very heterogeneous, and included several complex mixtures. Furthermore, the comparison did not take account of temporal factors such as latency and age at exposure, and other variables which may be of importance in the prediction of human cancer risk.

One group of carcinogens for which extensive human and experimental results are available consists of the drugs used in cancer therapy. Although often of considerable therapeutic value, these agents can also be viewed as models for human carcinogenesis. They can thus provide a framework within which a number of aspects of cancer risk quantification can be investigated, and indeed such investigation may lead to a reduction in the long-term hazards of cancer therapy. In the paper, we describe the quantitative information which is available concerning the carcinogenicity and related properties of the anticancer drugs, with particular emphasis on the mechanisms by which they may be exerting their carcinogenic activity.

2.   CHEMOTHERAPEUTIC AGENTS

At least 16 drugs known to have been used in cancer therapy have been shown to increase cancer risk in humans or experimental animals [14]. Most of these drugs are alkylating agents, which covalently bind to cellular DNA either directly or after metabolic activation. In humans, leukemia is the main cancer induced by alkylating agents, but experimental animals exhibit increases in a wider range of solid and haematopoietic tumours following exposure. Adriamycin and actinomycin D are the only non-alkylating drugs which have so far been identified as animal carcinogens. They have not been recognized as human carcinogens. Table 1 summarizes the identified target organs for carcinogenesis of some anticancer agents in humans, mice and rats.

Table 1:   Identified human and animal target organs for the carcinogenicity of selected anticancer agents [14].

Target Organs for cancer in:

Drug	Humans	Mouse	Rat
Cyclophosphamide	Haematopoietic system Bladder	Haematopoietic system Lung Mammary gland	Haematopoietic system Bladder Mammary gland
Melphalan	Haematopoietic system	Lung Lymphatic system	Mammary gland
Procarbazine	?	Haematopoietic system Nervous system Lung	Haematopoietic system Nervous system Mammary gland
Adriamycin	?	Mammary gland	
Actinomycin D	?	Local sarcomas[1]	Local sarcomas[1]
Cis-Platinum	?	Lung	Haematopoietic system Kidney

[1] At the site of administration

The use of anticancer drugs provides perhaps the only situation under which a number of factors related to chemical carcinogenesis can be effectively studied in humans. In particular, they are pure compounds, and the only human carcinogens to which humans are intentionally exposed at doses which, like in animal experiments, are close to the threshold of acute toxicity. The exposure levels are carefully measured and monitored, treated patients are routinely observed for long periods following treatment, and other concurrent biological measures are often made. Furthermore, the pharmacokinetic and metabolic properties of the drugs are intensively investigated because of their importance in the understanding of therapeutic efficacy. Finally, the acute leukemias which result from alkylating agent therapy in between 2 and 10% of 10-years survivors (depending on the agent and first cancer type) are very rare in the absence of chemotherapy, so that the background risk is by comparison negligible, and they are usually manifested within 10 years of treatment.

There are, on the other hand, several reasons why anticancer drugs may not be considered appropriate models for studying cancer risk quantification in general. Human exposure to high doses of alkylating agents is rare, and is probably responsible for only a small fraction of acute leukemias, let alone cancer in general. It may be that the mechanisms by which high-dose alkylating chemotherapy causes leukemia in humans is quite different from those involved in other, more common forms of cancer. Acute leukemias are rare forms of cancer comprising less than 1% of the total in most cancer registries [19], and are not of the epithelial type, the group to which the vast majority of human cancers belong. While certainly true, these arguments are not necessarily in opposition to the previously stated advantages of studying anticancer drugs as a model for human carcinogenesis. Indeed, the very specificity of the agents and types of cancer induced should facilitate the process of risk modelling and quantification, and at least provide a demonstration of risk estimation methodology under ideal conditions. The fact that alkylating agents do not appear to cause a substantial increase in solid tumours in humans may be of substantial interest in itself, since they are certainly capable of inducing carcinomas and other tumours in animal experiments. If real, this difference may shed light on some fundamental lack of comparability in the susceptibility of humans and the rodent species used for carcinogenicity testing.

Over the past several years, we have been conducting a series of studies of second cancer in relation to chemotherapy. The studies are carried out by a collaborative group of cancer registries and large hospitals, and have so far involved a cohort study and several case-control studies [15]. Their primary goal is the reduction of long-term risks for cancer survivors, but the studies are also likely to provide a considerable amount of information on the quantification of human cancer risk.

3. BUILDING A MATHEMATICAL MODEL OF ALKYLATING-AGENT LEUKEMOGENICITY

For the reasons outlined in the preceding section, anticancer drugs offer unique possibilities for studying quantitative aspects of carcinogenesis and, by corollary, for developing mathematical models to describe the carcinogenic process.

In its most general form, a mathematical carcinogenesis model specifies the risk of cancer for an individual, given any exposure history and set of personal characteristics. Models of this degree of generality can certainly be proposed on theoretical grounds, but they could probably never be fully validated, because of the limitations of epidemiological and other empirical information on human cancer risk. More practical model formulations restrict themselves to the description of risk following a limited range of exposure patterns, and at best take account of only the most general individual characteristics, such as age and sex. Predictions of such models can be more readily compared to available epidemiological data. Nevertheless, it is clear that even the most basic mathematical cancer model must in some way take account of several fundamental features of carcinogenesis. Of primary importance are:

(i)   the relationship between dose level and the
       degree of carcinogenic effect;
(ii)   the time course of the carcinogenic effect, often
       referred to as the latency;
(iii) the way in which exposures at different time
       periods combine to affect cancer risk.
Further elaborations of cancer models might include:
(iv)   the extent to which latency differs with exposure
       level;
(v)   the way in which personal factors and other
       exposures may affect (i)-(iv) above.

Mathematical models for the development of cancer in
humans are formulated in two general ways, each of which
has advantages and disadvantages. Either they are deduced
from a list of theoretical premises about the biological
processes involved, as has been the case with the
multistage [2] and two-stage [18] models, or they are
constructed empirically, starting from observational and
experimental carcinogenesis data and proceeding with linear
models and other standard tools of applied statistics.
There may of course be overlap and even convergence in the
two approaches, in that the models based on biological
theory are modified using real data, and biological
judgment may influence the choices which must be made when
carrying out statistical model-building. In the following
sections we take the empirical approach, in examining the
information which has so far accumulated on points (i)-
(iii) above, in relation to the leukemogenicity of the
alkylating agents used in cancer chemotherapy. In doing
so, however, the implications for the biological basis of
cancer models will be discussed where relevant.

Dose-effect relationships

Studies of patients treated with single-agent
chemotherapy are generally easier to interpret than those
in which multiple agents were used, and are particularly
valuable in the investigation of dose-effect relationships.

Kaldor, Day and Hemminki [16] reviewed studies of leukemia following treatment with single agents, or at most one alkylating agent. There are relatively few studies of this kind, and the degree of published detail on the relationship between dose and leukemia risk has been very limited. Most of the studies give no more than the cumulative or relative risk of leukemia over a specified time period, and the average total dose of drug received by the patients. These figures can be used to estimate leukemogenic potency, as the risk per unit total dose, provided assumptions are met concerning linearity of leukemia risk with dose, constant incidence with time, absence of a dose-rate effect and the role of the background risk. There is, however, no way to examine these assumptions for most of the published studies, either because the number of observed cases is simply too small, or because the data were not presented in a suitable way.

There are several published studies which give estimates of risk for several categories of total dose. In Figure 1, the excess relative risk (defined for each category as one less than the ratio between the risk of leukemia in the category and the risk among patients not treated at all with chemotherapy) is plotted against dose for melphalan [12], cyclophosphamide [13] and methyl-CCNU [4]. The dose assigned to each category is simply the midpoint of the interval (except for the highest category, in which it is the lower cutpoint of the interval), and in the Figure, the doses for each drug have been rescaled by dividing by the dose corresponding to the highest category for the drug. There is no suggestion of a threshold of carcinogenic effect. There does appear to be upward convexity in the curves, but it may be largely due to underestimation of the dose level in the highest group through the use of its lower cutpoint. Curvature of this kind, if real, would suggest that the drugs cause leukemia by affecting several steps in the pathway to malignancy [9].

FIGURE 1. Excess relative risk of leukemia in humans by
total dose, for selected alkylating agents

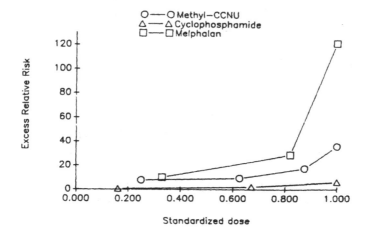

So far, no published study on leukemia following
chemotherapy has attempted a formal evaluation of the shape
of the dose-effect relationship. The shapes suggested by
Figure 1 may well be misleading if the dose-points chosen
to represent each category are not representative of the
true doses in the category. A more appropriate analysis
would plot the excess relative risks against the mean or
median dose in each category and take account of age or
other differences among the dose groups.

The differences in excess relative risk among the
three drugs are impressively consistent across the levels
of dose. Even after taking account of the absolute doses
used by internally scaling on the basis of the highest dose
category used, cyclophosmadie was by far the weakest
leukemogen of the three.

The quality of information on dose-response shape from
animal cancer tests of cytotoxic drugs has also been
variable. In publications summarizing studies which
involved long term administration of test compounds [11],
estimates of carcinogenic potency are presented, as well as

other information concerning the bioassay results. Upward curvature in the dose-response curve, as evaluated by a statistical hypothesis test, has been noted for procarbazine, thio-TEPA and isophosphamide.

Latency of leukemogenic effect

The temporal pattern of chemotherapy administration has varied substantially, according to the type of cancer treated, the drugs used and the occurrence of relapse in individual patients. Evaluation of the latency of the leukemogenic effect is complicated by this variation, particularly for chemotherapy given over extended time periods. In this situation, the latency time scale becomes confounded with the effect of duration of exposure, just as it does in studies of cancer risk in industrial and other cohort studies [7]. Considerable simplification can be achieved by considering only patients who had a single episode or course of chemotherapy in a limited time period, just as estimation of dose-effect can be most straightforwardly made for patients who were only treated with one drug. Again, however, there is little relevant published information available. Several studies have presented the cumulative incidence (Kaplan-Meier) curve for leukemia as a function of time since the first cancer. This curve certainly provides essential descriptive information on the evolving risk of leukemia, but it is only of limited use in establishing the latency function for the effect of chemotherapy on the risk. As usually presented, the cumulative incidence curve does not distinguish among different temporal patterns of treatment, although in some papers, it has been presented separately for broad treatment categories. Furthermore, it does not take account of the evolution of the background cancer risk with time. Although this risk may be negligible compared to the risk in patients treated with chemotherapy, it increases substantially with age, and is therefore of importance when comparing the evolution of risk among groups of patients treated at different ages.

Some studies [12, 6] have reported the risk of leukemia as a function of time since the first chemotherapy, either in absolute terms or relative to the risk in patients never treated with chemotherapy. It is often easier to discern patterns in risk estimates broken down in this way than in the cumulative incidence curves, which consist of steps at the time of incidence of each new case and are by definition always increasing. The overall conclusion from the ensemble of studies which report on incidence by time since first chemotherapy is that the risk increases substantially within one or two years, peaks at about 5 or 6 years both in relative and absolute terms and declines thereafter. There is not yet enough accumulated observation time to evaluate accurately the risk past about 10 years, and in particular to know whether and when it returns to the level in patients never treated with chemotherapy.

Cuzick et al. [8] have tried to identify the period of greatest risk following treatment, by considering treatment in specific time intervals prior to the onset of leukemia. They conclude that it is the most recent three-year period which is of greatest importance in determining subsequent leukemia risk. The period was identified by examining the likelihood as a function both of selected intervals, and a parameter relating risk to the total dose received in the intervals. When applied formally, this procedure results in maximum likelihood estimates of the interval and the relationship between risk and dose. The two estimates may well be rather highly correlated, since the shape of the dose-risk relationship may be largely dependent on the interval in which the dose is accumulated. Figure 2 shows the results of estimating the interval in this way, using data from a case-control study of second cancer. The relative risk of second cancer was estimated as a log-linear function of total dose of chemotherapy received, as treatment for the first cancer, only taking into account

the chemotherapy received in intervals which either
included or excluded the number of most recent years
indicated on the x-axis. The conclusions reached from this
analysis would be somewhat different from those of Cuzick
and colleagues. Certainly the treatment in the most recent
years is the greatest determinant of risk, but it would
appear that substantial improvement in the likelihood is
obtained by including treatment up to 7 years before the
onset of leukemia.

Figure 2.   Likelihood estimation of the interval during
             which chemotherapy has an effect on leukemia
             risk.

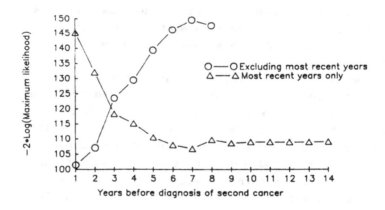

Even latency functions with this form are
oversimplifications [17] and could be elaborated by the use
of smoother functional forms which allow for more gradual
changes in the latency effect with time. Thomas [22] has
used a lognormal density function for this purpose in
studying lung cancer risk among asbestos workers.

The combined effect of several treatment episodes

A substantial proportion of cancer patients receive chemotherapy on several occasions or over an extended time period, either as treatment for relapse of the original tumour or as adjuvant therapy following its surgical removal. Estimation of leukemia risk in this situation requires an understanding of the way in which the effects of the episodes combine. Under the simplest model, they would be expected to combine additively, increasing the risk by effectively overlaying the dose and latency effects from the multiple episodes. The validity of this model can be investigated by evaluating the risk in patients among whom one subgroup had received chemotherapy only over a short period, and the rest had been subsequently retreated with chemotherapy. Although the parameters of the model for the effect of multiple episodes could be estimated without the first subgroup of patients, this group provides an essential check of the estimated latency function's validity.

An analysis of this kind has so far not been undertaken for patients treated by chemotherapy. In another paper in this volume (Grosser and Whittemore), the effect of multiple exposure episodes is examined for experimentally induced tumours in animals.

Simultaneous estimation of effects

So far, we have attempted to describe various aspects of chemotherapy-induced leukemia risk in a way which is, as far as possible, not linked to any mathematical model. By considering groups of patients who are more or less comparable apart from the particular risk determinant whose effect is being evaluated, it is possible to avoid making assumptions about the functional from of the relationships being estimated. The penalty for choosing such selected groups of patients is that the number of cases in each category of the variable under study is reduced and the

estimation precision is correspondingly decreased. This
problem is compounded when one turns to the simultaneous
estimation of effects.

A full description of the leukemogenic risk due to
chemotherapy which is not dependent on a mathematical model
would require risk estimates simultaneously broken down by
age at treatment, drug type and dose, temporal pattern of
treatment, latency and individual characteristics such as
age and sex. Since even in a big study the sample size is
rapidly exhausted by multiple cross-classifications of such
explanatory variables, one is sooner or later obliged to
adopt a parametric model of the leukemia risk. For
example, in a recent paper [17], we proposed a bilinear
form to describe the effect on second cancer risk of drug
type, dose and latency. The model had the form:

$$R = \sum_{ij} \alpha_i \, \beta_j \, d_{ij} \quad ,$$

in which the leukemia risk R is expressed as the sum of
terms which are the product of the doses d_{ij} of the ith
drug received in the jth time period before leukemia, and
parameters α_i and β_j (to be estimated) which correspond to
the effects of different drugs and latency periods,
respectively. The assumption behind this model is that the
latency effect has the same general shape for all drugs
under consideration, and simply varies by a constant amount
for each drug.

4.   OTHER ASPECTS OF CHEMOTHERAPY INDUCED CANCER

In addition to providing a basis for developing a
mathematical model to represent the relationship between
chemotherapy and the subsequent risk of leukemia, studies
of medically induced cancer can yield other kinds of
information which would be difficult to obtain for other
human carcinogens. Some areas of particular interest are
breifly discussed below.

Comparison with radiation carcinogenesis

Ionizing radiation induces tumours in a wide range of human tissue types and organs, and in particular, is capable of causing acute non-lymphocytic leukemia. In its lack of organ specificity, it appears to differ from alkylating chemotherapeutic agents, which so far have been only demonstrated to cause leukemia in humans, with the exception of cyclophosphamide, which is also a bladder carcinogen.

For a number of forms of cancer, both radiotherapy and chemotherapy are used, either alone or in combination. Studies of leukemia risk among survivors allow a comparison of leukemogenic potential to be made between radiotherapy and chemotherapy, and in particular provide the basis for attempting a calibration of the two types of leukemogens. Dose of radiation can be estimated for the bone marrow [5], the presumed target organ for leukemogenesis, and the increase in leukemia risk per unit dose compared with the increase due to chemotherapy. Most studies of chemotherapy-related leukemia published to date do not provide radiation dose estimates, but it is likely that in the future, case-control investigations of leukemia will yield information of this kind.

Investigation of the chemotherapy dose distribution in the body could be used as the basis for predicting the increase in the risk of cancers other than leukemia following chemotherapy, on the assumption that proportionability is maintained between the effects due to chemotherapy and radiation. If indeed the risk for solid tumours is substantially lower than predicted by this assumption, serious doubt would be cast on the use of concepts such as the "rad equivalent" [10] to standardize cancer risk estimation for chemicals.

Precursors to alkylating agent induced cancer

Of particular importance in the mathematical modelling of carcinogenesis is the degree to which information on intermediate stages and precancerous states is available [18]. Although there is a well defined condition known as the myelodyplastic syndrome [3] which can precede acute leukemia following alkylating agent therapy, it does not always lead to leukemia, and can in fact be fatal in itself. It is well established that chemotherapy related leukemias contain characteristic chromosomal deletions [20] and in the future their presence may offer another preleukemic state to study, if it is possible to reliably detect them in single cells which have not yet undergone clonal expansion.

A third possibility is the degree of cytopaenia following therapy, which may be an indicator of susceptibility to DNA alkylation or repair status, which in turn may be correlated with leukemogenesis. Of considerable interest would be analyses of the relative risk of leukemia by category of cytopaenia in various time periods during and following the cessation of chemotherapy for Hodgkin's disease or some other malignancy. A finding that, for a given amount of chemotherapy, white cell counts are predictive of leukemia risk, would suggest either an association between bone marrow sensitivity and leukemia susceptibility, or an intermediate role for bone marrow proliferation in the development of leukemia.

Predictive value of in vitro tests and animal cancer bioassays

Since the review described above [16] there have not been reports of further animal cancer tests of alkylating drugs which could supplement the study of potency correlation between animals and humans. The most extensive bioassays have been those carried out by the U.S. National Toxicology Program, utilizing chronic lifetime exposure to the test chemicals. While perfectly appropriate for carcinogen testing, this type of study is obviously rather different from the situation of patients undergoing cancer therapy, who generally receive a high dose for a very short fraction of their lives. Furthermore, the route of exposure in many animal studies has been intraperitoneal, which certainly results in whole-body exposure but may also produce tumours of rather specific types locally at the site of injection.

Even though the carcinogenicity of a number of chemotherapeutic agents has been well established in animal experimentation, it would clearly be of value to carry out further experiments which mimic more closely the human exposure situation, and which would also provide the possibility of carrying out analyses of DNA adducts, chromosomal aberrations and other short-term end points which are now being extensively studied in human subjects. In this way, the validity of the animal model could be evaluated for several components of the leukemogenesis process. More clinically oriented bioassays have already been implemented in this way [21] but they remain the exception overall.

In vitro short-term tests and pharmacokinetic studies of alkylating and other anticancer drugs could also be usefully repeated, with a view to understanding the intermediate steps between treatment with alkylating drugs and the long-term consequences. Some drugs, such as adriamycin and cisplatinum, are clearly carcinogenic to animals, but may not be in humans. Explanations of these differences could also shed considerable light on species differences in metabolism, DNA repair or other factors.

5.   CONCLUSION

Because such a small and specific subgroup of the population is exposed to anticancer agents, they are rarely considered among the important human carcinogens. However, the fact remains that we could potentially know more about their mechanism of carcinogenic effect than we could about any other group of agents. A considerable amount of information has already accumulated, mostly as a consequence of research into improving the drugs' therapeutic effectiveness. While such improvement is clearly the primary objective of research on the anticancer agents, it should be possible to gain substantial insight into the process of modelling human carcinogenesis along the way.

REFERENCES

1. B.C. ALLEN, K.S. CRUMP and A.M. SHIPP. Correlation between carcinogenic potency of chemicals in animals and humans. Risk Analysis, 8 (1988), pp. 531-544.

2. P. ARMITAGE and R. DOLL. The age distribution of cancer and a multistage theory of carcinogenesis. British Journal of Cancer, 8 (1959), pp. 1-12.

3. J.M. BENNETT, D. CATOVSKY, M.T. DANIEL et al. The morphological classification of acute lymphoblastic leukaemia: Concordance among observers and clinical correlations. British Journal of Haematology, 47 (1981), pp. 553-561.

4. D.W. BLAYNEY, D.L. LONGO, R.C. YOUNG, M.H. GREENE, S.M. HUBBARD, M.G. POSTAL et al. Decreasing risk of leukemia with prolonged follow-up after chemotherapy and radiotherapy for Hodgkin's disease. New England Journal of Medicine, 316 (1987), pp. 710-714.

5. J.D. BOICE Jr., M.H. GREENE, J.Y. KILLEN Jr., S.S. ELLENBERG and J.G. FRAUMENI Jr. Leukemia after adjuvant chemotherapy with semustine (methyl-CCNU). Evidence of a dose-response effect. New England Journal of Medicine, 314 (1986), pp. 119-120.

6. J.D. BOICE, M. BLETTNER, R.A. KLEINERMAN et al. Radiation dose and leukaemia risk in patients treated for cancer of the cervix. Journal of the National Cancer Institute, 74 (1987), pp. 1295-1311.

7. N.E. BRESLOW and N.E. DAY. Statistical Methods in Cancer Research, Volume II, The Design and Analysis of Cohort Studies (IARC Scientific Publications No. 82), Lyon, (1987) International Agency for Research on Cancer.

8. J. CUZICK, S. ERSKINE, D. EDELMAN and D.A.G. GALTON. A comparison of the incidence of the myelodysplastic syndrome and acute myeloid leukaemia following melphalan and cyclophosphamide treatment for myelomatosis. British Journal of Cancer, 55 (1987), pp. 523-529.

9.   N.E. DAY and C.C. BROWN. Multistage models and the
     primary prevention of cancer. Journal of the
     National Cancer Institute, 64 (1980), pp. 977-989.

10. L. EHRENBERG, E. MOUSTACCHI, S. OSTERMAN-GOLKAR.
     Dosimetry of genotoxic agents and dose-response
     relationships and their effects. Mutation
     Research, 123 (1983), pp. 121-182.

11. L. GOLD, C. SAWYER, R. MAGAW, G. BACKMAN, M. DE
     VECIANA, R. LEVINSON, N. HOOPER, W. HAVENDER, L.
     BERSTEIN, R. PETO, M. PIKE and B. AMES. A
     carcinogenic potency data base of the standardized
     results of animal bioassays. Environmental Health
     Perspectives, 58 (1984), pp. 9-319.

12. M.H. GREENE, E.L. HARRIS, D.M. GERESHENSON, G.D.
     MALKASIAN, L.J. MELTON, A.J. DEMBO et al.
     Melphalan may be a more potent leukemogen than
     cyclophosphamide. Annals of Internal Medicine, 105
     (1986), pp. 360-367.

13. J.F. HAAS, B. KITELMAN, W.H. MEHNERT, W. STANECZEK,
     M. MOHNER, J. KALDOR and N.E. DAY. Risk of
     leukaemia in ovarian tumour and breast cancer
     patients following treatment by cyclophosphamide.
     British Journal of Cancer, 55 (1987), pp. 213-218.

14. IARC. Overall Evaluations of Carcinogenicity: An
     Updating of IARC Monographs Volumes 1 to 42. IARC
     Monographs on the Evaluation of Carcinogenic Risks
     to Humans, Supplement 7 (1987), International
     Agency for Research on Cancer, Lyon, France.

15. J. KALDOR, N.E. DAY, P. BAND, N.W. CHOI, E.A.
     CLARKE, M.P. COLEMAN et al. Second malignancies
     following testicular cancer, ovarian cancer and
     Hodgkin's disease: An international collaborative
     study among cancer registries. International
     Journal of Cancer, 39 (1987), pp. 571-585.

16. J. KALDOR, N.E. DAY and H. HEMMINKI. Quantifying
     the carcinogenicity of antineoplastic drugs.
     European Journal of Cancer Clinical Oncology, 24
     (1988), pp. 703-711.

17. J. KALDOR and N.E. DAY. Estimation of temporal
     effects in treatment-induced second cancer.
     Statistics in Medicine, 8 (1989) (in press).

18. S.H. MOOLGAVKAR and A.G. KNUDSON Jr. Mutation and
     Cancer: A model for human carcinogenesis. Journal

of the National Cancer Institute, 66 (1981), pp. 1037–1051.

19. C.S. MUIR, J. WATERHOUSE, T. MACK, J. POWELL and S. WHELAN. Cancer Incidence in Five Continents, Volume V. (1987) IARC Scientific Publications No. 88.

20. J.D. ROWLEY, H.M. GOLOMB and J.W. VARDIMAN. Nonrandom chromosome abnormalities in acute leukemia and dysmyelopoietic syndromes in patients with previously treated malignant disease. Blood, 58 (1981), pp. 759–767.

21. D. SCHMAHL and H. OSSWALD. Experimental studies on the carcinogenic effect of cancer chemotherapeutic agents and immunosuppression. Arzneimittelforschung, 20 (1970), pp. 1461–1467.

22. D.C. THOMAS. Statistical methods for analyzing effects of temporal patterns of exposure on cancer risks. Scandinavian Journal of Work and Environmental Health, 9 (1983), pp. 353–366.

AFTERWORD: SOME THOUGHTS ON WHAT WAS LEARNED AND SOME SCIENCE POLICY ISSUES

James D. Wilson

ABSTRACT

This symposium marked a significant change in the use of mathematical modelling to understand carcinogenesis. Within the last two years, experimentalists in carcinogenesis have begun to recognize the power for testing hypotheses offered by expressions of the new theory first described by Moolgavkar and Knudson in 1981. We have already seen significant change in our understanding as a result.

Any new field of scientific investigation develops its own language adapted to communicate concepts peculiar to its investigators' needs. This new field of carcinogenesis modelling borrowed from pathology the concept of "stage" to describe discrete steps in the process by which normal cells are transformed into cancer cells. Because its use evolved, communication between mathematicians and biologists became difficult: the mathematical models became focussed on the events which cause transition from one biological "stage" to the next, and the usage changed as a result. We suggest that modellers adopt "event" instead of "stage" to describe their expressions, in order to clarify communications.

The models explored here are potentially useful for regulatory purposes — sometimes referred to as "risk assessment". Their use for hypothesis-testing is clear. However, it seems likely that unavailability of data will preclude widespread use of these models to predict hazard functions far outside the observable range; that is, their direct usage for regulation will not soon be common. In addition, the uncertainty associated with estimating all the parameters which enter these calculations means that the final estimates of the hazard function cannot be very precise. A means is needed for altering the present regulatory methodology so as to take into account the

information developed by application of the models
discussed at this symposium to the experimental data whose
generation they suggest.

I. INTRODUCTION

The Symposium translated here into print could be
considered the second in an irregular series between mathematicians interested in modelling carcinogenesis and
progression, and experimental biologists who can provide
the data those modellers need. The first of these was held
in Corfu, Greece, in June, 1988, under the sponsorship of
NATO and the United States National Science Foundation; the
proceedings have been published[7]. As happens when any
line of investigation turns into a new field of research,
communication has not always been smooth. This is to be
expected, especially, of a new field like this one, emerg-
ing out of two established and very different fields. The
Corfu symposium began the process of defining this new
field, and identified both some of its problems and opportunities. This Snowbird symposium made great strides toward putting these problems and opportunities in perspective for people from both the established fields, statistical modelling and experimental carcinogenesis. It also
provided both groups with a much better appreciation of the
concepts which underlie cancer modelling, and give it the
power it has to test hypotheses. It was both exhausting
and exhilarating.

II. THE CONCEPT AND SEMANTICS OF "STAGE"

One key concept that proved to be understood differently by the two groups was that of "stage". That misun-
derstanding can be resolved — and was for those present —
by having mathematicians adopt a more accurate term; we
suggest "event". Cancer biologists and statistical modellers use it differently, and that difference has impeded the
modellers' ability to impart to the biologists both the
power of their tools and the validity of the conclusions
drawn through use of those tools. To the biologists
"stage" denotes a cellular mass with visually distinguishable characteristics, regarded as an identifiable way-
station in the continuous process through which normal
cells evolve into invasive, metastatic cancerous tumors.
(A useful analogy might be the "equilibrium" parts of the

"punctuated equilibrium" theory of paleontological evolution.) It seems likely that this term may have been borrowed by the experimental biologists from their counterparts in clinical oncology: there the term is used to
characterize tumors and communicate about prognosis. The
usage is similar.

It also seems likely that its meaning to mathematicians evolved after its initial usage in the 1950s. Armitage and Doll[1] clearly used the term in the same way as
experimentalists do now. The probability that a tumor
would appear depended on the length of time spent travers-
ing each stage; carcinogenic agents were seen as speeding
that traverse.

However, in the interim, the theory enunciated by
those authors evolved. As biochemical genetics matured in
the 1950s and 1960s, the concept of mutation changed: attention focussed on the reaction of chemical mutagens with
DNA, with the resulting DNA adduct being responsible for
misreading of bases as duplication proceeded during mitosis. The rate at which such reactions occur was apparently
assumed to limit the progress through Armitage and Doll's
"stages"; this mutational "event" became identified with
the "stage". Mathematically, this makes no difference; the
speed of the traverse is determined by the rate at which
events causing transitions from one "stage" to the next
occur. However, the use of "stage" (e.g., by Crump et
al.[2]) instead of "event", when the conceptual model on
which the mathematics is built clearly is more accurately
described by the latter, confuses the biologists.

This possibility for confusion was noted, perhaps
implicitly, by Moolgavkar and Knudson; their 1981 paper[5]
distinguishes the pathologically identifiable "stages" from
the "events" — now specifically identified as rare mutations — which cause transitions from one stage to the next.
However, that distinction has not always been kept clear
(cf. [4,6]).

The mathematical models now being explored — the
subject-matter of this symposium — now focus on the rate-
limiting (or, better, probability-determining[3,4]) events,
and not the oncologic stages. Because this concept of
rate-limiting events is not part of the common intellectual
armamentarium of experimental biology, modellers have difficulty communicating to biologists how use of models built

on it can be used to test hypotheses. Thus we should re-
serve "stage" for referring to the intervals between the
transitional "events" critical to the models explored here.

III. UNCERTAINTY IS TOO GREAT TO ALLOW USE OF A STATISTI-
CALLY-DEFINED UPPER CONFIDENCE LIMIT ON THE HAZARD FUNCTION
FOR STANDARD-SETTING

This Symposium's title implied a focus on use of
carcinogenesis modelling in risk assessment. The discussions suggested some difficulties in making such use of the
models, and that in turn raises some issues of science
policy. In particular, it seems clear that limitations on
the data available ensures that very rarely will the model
outputs be used to estimate low-exposure responses. Yet it
also seems clear that use of the models yields information
about those responses which policy makers should find very
useful. So the question becomes, how to make that information available in the regulatory context.

To explore that, a few words about the regulatory
process may be useful. It almost never involves estimation
of risk, *per se*. The objective of the regulator is prevention of injury to those members of the public exposed to
the chemicals he or she regulates. This end may be
achieved by several means: setting an enforceable toler-
ance limit or exposure standard, awarding a license for
certain uses, or (its inverse) banning certain uses, etc.
In this decade we have come to call the science-based process which supports these regulatory actions "risk assessment" — influenced, presumably, by the National Academy of
Science's " red book" *Risk Assessment in the Federal
Government: Managing the Process* (National Academy Press,
1983).

Generally this "risk assessment" process is carried
out by public health professionals, frequently toxicolo-
gists. Toxicology is a relatively young profession,
roughly half a century old (although the "science of poi-
sons" goes back many centuries). One thing which charac-
terizes modern toxicology is a heavy use of data from studies in animals as predictors of responses in humans. The
experts in this field are keenly aware of the limitations
imposed by this reliance on animals; both qualitative and
quantitative uncertainties abound. From the cumulative
experience of the profession has arisen a series of prac-

tices — guideposts — used to help set standards or otherwise regulate exposure in the face of these uncertainties. One now well-established practice is the use of the "plausible upper bound" on risk.

The "upper bound" part of this phrase makes it sound as though it were a statistical quantity. Really it isn't. It may have been statistical in origin, but as it was described by Dr. Daniel Byrd in a comment during the last discussion period of this symposium, it lost its strict statistical meaning some time ago. Dr. Byrd characterized the "plausible upper bound" as including substantial elements of judgment mixed into the relatively simple mathematics. To him, the "linearized multistage" procedures provide a means of arriving at an exposure "safe enough" for regulatory purposes.

As I have pointed out elsewhere[8], these linear low-dose procedures derive from a theory of carcinogenesis that is now obsolete. As the results described in this symposium make clear, increased mitotic rate is a significant risk factor for carcinogenesis, and mitogenic stimuli interact synergistically with exposures which cause mutations. These procedures do not allow consideration of data which indicate that significant nonlinearities may be resulting from this synergistic interaction, and thus significantly misdescribe the dose-response. (The synergism can be seen in the "hockey stick"-shaped dose-response curves found for formaldehyde in the rat nose, described by Starr, and for 2-AAF in the mouse bladder, described by Cohen; consult their papers in this symposium.)

In principle, development of the models discussed at this symposium and the data needed to apply them, so that they could routinely be applied, would provide a way to obtain consideration of relevant data in the regulatory process. In practice, that will prove difficult, if not impossible. One difficulty lies in the amount of data required to implement these models: data on the dose-response of net birth rate of cells from an initiated population is not commonly available, for instance. Another serious difficulty was raised by Chris Portier's paper (q.v.) on the effect of multiple uncertainties on estimated "risk". Portier showed that if the several parameters which enter the models are considered to be distributions, the distribution of risk resulting from their propagation is very broad. In another paper at this symposium, Curtis

Harris described several souces of human variability that are reflected in a distribution of susceptibility to cancer, including different enzyme activities and DNA repair capacities. These do not enter directly into the modelling efforts discussed here, but do require that the input parameters realistically be considered as distributions.

Portier presented a paradox. He compared the uncertainty in two estimated hazard functions, one carried out using conventional default assumptions, the other using a modelled estimation of the conversion from incident exposure in animals to target-tissue dose in man. Because the latter includes many parameters which must separately be estimated and which thereby must be considered to be distributions (because of measurement error as well as inherent variability), the propagated uncertainty results in a substantial uncertainty in the final estimate of hazard at any incident exposure, larger than that derived for the conventional method. Most scientists expert in this field would believe that the modelled hazard is more accurate than the conventional estimate, and have been surprised that use of this more accurate estimate could lead to an increase in the "upper bound". The source of this apparent paradox is clear: use of the default procedure does not yield an estimate of the 95% upper confidence limit of the hazard, as Portier defined it.

This observation is not inconsistent with Byrd's description of agency "risk assessment" practice; neither is it inconsistent with official pronouncements. (FDA, for instance, has always recognized that its standards will not necessarily protect everyone.) The standards are the result of expert weighing of the uncertainties to arrive at a what the experts' judgement regards as safe enough, and not just the result of statistical evaluation of the data amenable to such treatment.

Byrd's remarks tell us that regulators need relatively simple methods for standard-setting. Good policy requires that these methods be able to incoporate all the reliable information relevant to the determination. We in the regulatory science business need to address how to evolve current practices to allow incorporation of the scientific information produced by these models. Not enough science is yet available to do this, but the results described in this Symposium give confidence that it will not be too long before the process can begin.

312

IV. REFERENCES

1) P. Armitage and R. Doll, "The age distribution of cancer and a multi-stage theory of carcinogenesis". *Brit. J. Cancer* **8**:1-12 (1954).

2) K. S. Crump, D. G. Hoel, C. H. Langley and R. Peto, "Fundamental carcinogenic processes and their implications for low dose risk assessment".*Cancer Research* **36**:2673-2679 (1976).

3) R. E. Greenfield, L. B. Ellwein and S. M. Cohen, "A general probabilistic model for carcinogenesis: analysis of experimental urinary bladder cancer". *Carcinogenesis* **5**:437-445 (1984).

4) S. H. Moolgavkar, A. Dewanji, and D. J. Venzon, "A stochastic two-stage model for cancer risk assessment. I. The hazard function and the probability of tumor." *Risk Anal.* **8**:383-392 (1988).

5) S. H. Moolgavkar and A. G. Knudson, Jr., "Mutation and cancer: a model for human carcinogenesis". *J. Nat. Cancer Inst.* **66**:1037-1052 (1981).

6) T. W. Thorslund, C. C. Brown and G. Charnley, "Biologically motivated cancer risk models". *Risk Anal.* **7**:109-119 (1987).

7) C. C. Travis, ed., Biologically Based Methods for Cancer Risk Assessment. New York and London: Plenum Press (in cooperation with NATO Scientific Affairs Division), 1989.

8) J. D. Wilson, "Biological bases for cancer dose-response extrapolation procedures". *Env. Health Perspectives*, (in press).